WATER RESOURCE PLANNING, DEVELOPMENT AND MANAGEMENT

WATER UTILITIES USE OF ASSET MANAGEMENT PRACTICES

ASSESSMENTS AND OPPORTUNITIES

WATER RESOURCE PLANNING, DEVELOPMENT AND MANAGEMENT

Additional books in this series can be found on Nova's website under the Series tab.

Additional e-books in this series can be found on Nova's website under the eBooks tab.

WATER RESOURCE PLANNING, DEVELOPMENT AND MANAGEMENT

WATER UTILITIES USE OF ASSET MANAGEMENT PRACTICES

ASSESSMENTS AND OPPORTUNITIES

ALFRED CALDWELL
EDITOR

NOVINKA

New York

Library of Congress Cataloging-in-Publication Data

ISBN: 978-1-63485-937-0

Published by Nova Science Publishers, Inc. † New York

CONTENTS

PREFACE

Recent catastrophic breaks in water mains and sewer discharges during storms are indicators of the nation's old and deteriorating water and wastewater infrastructure. EPA estimates that small water utilities—those serving fewer than 10,000 people--may need about $143 billion for drinking water and wastewater infrastructure repairs and replacement over 20 years. EPA and USDA provide the three largest sources of federal funding for water infrastructure. In a March 2004 report, the United States Government Accountability Office (GAO) found that water utilities may benefit from implementing asset management—a tool used across a variety of sectors to manage physical assets, such as roads and buildings. This book examines what is known about the use of asset management among the nation's water utilities—particularly small water utilities— including benefits and challenges and steps EPA and USDA are taking to help small water utilities implement asset management.

In: Water Utilities Use …
Editor: Alfred Caldwell

ISBN: 978-1-63485-937-0

Chapter 1

WATER INFRASTRUCTURE: EPA AND USDA ARE HELPING SMALL WATER UTILITIES WITH ASSET MANAGEMENT; OPPORTUNITIES EXIST TO BETTER TRACK RESULTS*

United States Government Accountability Office

WHY GAO DID THIS STUDY

Recent catastrophic breaks in water mains and sewer discharges during storms are indicators of the nation's old and deteriorating water and wastewater infrastructure. EPA estimates that small water utilities— those serving fewer than 10,000 people--may need about $143 billion for drinking water and wastewater infrastructure repairs and replacement over 20 years. EPA and USDA provide the three largest sources of federal funding for water infrastructure. In a March 2004 report, GAO found that water utilities may benefit from implementing asset management— a tool used across a variety of sectors to manage physical assets, such as roads and buildings.

GAO was asked to review water utilities' use of asset management. This report examines (1) what is known about the use of asset management among

* This is an edited, reformatted and augmented version of a United States Government Accountability Office, Publication No. GAO-16-237, dated January 2016.

the nation's water utilities— particularly small water utilities— including benefits and challenges and (2) steps EPA and USDA are taking to help small water utilities implement asset management. GAO selected a nongeneralizable sample of 25 water utilities in 10 states based on largest infrastructure needs and interviewed EPA, USDA, state, and water utility officials.

WHAT GAO RECOMMENDS

GAO recommends that EPA consider collecting information about utilities' use of asset management through its needs assessment surveys, and that EPA and USDA compile the benefits of asset management into one document. EPA and USDA generally agreed with GAO's findings and recommendations.

WHAT GAO FOUND

The small water utilities GAO reviewed in 10 selected states are implementing some asset management practices, although state officials said that large water utilities are more likely to implement asset management than small utilities. The asset management practices these small utilities used include identifying key assets, such as pipelines, treatment plants, and other facilities, and assessing their life-cycle costs. For example, officials from 23 of the 25 small water utilities GAO reviewed said they had maps that identify the location of at least some of their assets. However, of the 25 small water utilities, officials from 9 said they knew the cost of rehabilitation versus replacement for all of their assets. Officials from the Environmental Protection Agency (EPA), U.S. Department of Agriculture (USDA), and the 10 selected states identified benefits and challenges for small water utilities using asset management. The benefits that EPA, USDA, and state officials identified include cost savings and more efficient long-term planning. The key challenges these officials identified include the availability of funding to cover start-up and maintenance costs, the availability of human resources, information on how to implement asset management practices, and political support from elected officials to begin an asset management program or increase user rates.

EPA and USDA are taking steps to help water utilities implement asset management by providing funding, free or low-cost tools such as software, one-on-one technical assistance, and classroom training for small water utilities that plan to implement asset management practices. EPA and USDA collect feedback from training participants, but do not collect information that will help track the results of the agencies' training efforts (e.g., whether utilities participating in such training implemented asset management practices). GAO identified in a March 2004 guide that evaluating training programs is key to ensuring training is effective in contributing to the accomplishment of agency goals and objectives. EPA officials told GAO that they had considered collecting nationwide data on water utilities' use of asset management but did not have the resources to pursue it. Leveraging existing data collection methods may be a cost-effective way for the agencies to collect this information. EPA conducts periodic needs assessment surveys of water utilities and has included questions about asset management use in the wastewater survey, but not in the drinking water survey. EPA officials said they did not receive enough responses to questions in the wastewater survey, and they have not considered including them in the drinking water survey. By continuing to include questions on wastewater utilities and considering questions about drinking water utilities' use of asset management in the surveys, EPA could have better assurance that it has information on the effectiveness of its training efforts with USDA. In addition, EPA and USDA officials told GAO that the agencies share anecdotal data on the benefits of asset management through technical assistance, but had not considered compiling such information into one document to encourage water utilities to adopt it. EPA and USDA are not required to compile such information, but doing so could provide information on benefits, including cost savings, and costs to water utilities that have not received training and could help encourage them to adopt asset management practices.

ABBREVIATIONS

CBO	Congressional Budget Office
CUPSS	Check Up Program for Small Systems
EFC	Environmental Finance Center
EPA	Environmental Protection Agency
GIS	geographic information system
ISO	International Organization for Standardization

OMB	Office of Management and Budget
PAS	Publicly Available Specification
SDWA	Safe Drinking Water Act
SRF	State Revolving Fund
USDA	U.S. Department of Agriculture

* * *

January 27, 2016

The Honorable Ken Calvert
Chairman
Subcommittee on Interior, Environment, and Related Agencies
Committee on Appropriations
House of Representatives

The Honorable Betty McCollum
Ranking Member
Subcommittee on Interior, Environment, and Related Agencies
Committee on Appropriations
House of Representatives

The Honorable Mike Simpson
House of Representatives

Recent catastrophic breaks in water mains and sewer backups and discharges during storms are indicators of the nation's old and deteriorating water and wastewater infrastructure. The American Society of Civil Engineers estimated in a 2013 report that there were more than 1 million miles of drinking water mains and 700,000 to 800,000 miles of public sewer in the country, much of which was in poor condition.[1] Drinking water and wastewater infrastructure that is old and deteriorating is at a higher risk of failure. Such infrastructure failures can disrupt the availability of drinking water and impede emergency response, such as fire-control efforts, or jeopardize public health by releasing sewage into surface waters and drinking water sources. Emergency repairs of such infrastructure can pose significant financial costs to communities. A water utilities' management professional estimated that emergency repairs can cost three to four times more than regular repairs—for example, if the cost to repair 1 mile of pipe is $550,000, the cost to repair this same pipe in an emergency could range from $1.7 million to $2.2 million.

According to the Environmental Protection Agency's (EPA) most recent estimates, more than $655 billion may be needed to repair and replace drinking water and wastewater infrastructure nationwide over the next 20 years.[2] Large water utilities—those serving populations of 100,000 or more—account for more of the projected infrastructure repair and replacement needs. EPA estimates that large water utilities may need an estimated $145 billion to repair and replace drinking water infrastructure and an estimated $219 billion for wastewater infrastructure (for a total of $364 billion). However, the majority of water utilities are small—serving populations of 10,000 or less,[3] and EPA estimates that small water utilities may need an estimated $110 billion for drinking water infrastructure and $33 billion for wastewater infrastructure (for a total of $143 billion).

Across the country, there are about 52,000 drinking water and 16,000 wastewater utilities. These water utilities[4] are generally subject to requirements under the Safe Drinking Water Act and Clean Water Act, respectively.[5] EPA and states regulate water utilities' compliance with these provisions, but the water utilities and the communities they serve are responsible for managing and funding the infrastructure needed to implement requirements, including the costs of repairs and replacement.

Communities can fund construction, repair, and replacement of infrastructure through such options as cash, municipal bonds, grants, and loans. The costs, including the repayment of loans, are funded through the rates communities pay on their water bills.

The federal government provides significant funding to assist water utilities. The three largest federally funded drinking water and wastewater infrastructure assistance programs are EPA's Drinking Water State Revolving Fund (SRF) program[6] and Clean Water SRF program,[7] and the Department of Agriculture's (USDA) Water and Waste Disposal program.[8] Through the Clean Water and Drinking Water SRF programs, EPA provides annual grants to state-level SRF programs. States use the SRF funds to, among other things, make loans to local communities and utilities for various water infrastructure projects, such as upgrading and replacing drinking water filtration and treatment plants and water mains, and constructing, repairing, and replacing sewage treatment plants and sewer pipelines. Communities of any size can apply for assistance through the SRF programs. In fiscal year 2015, EPA provided an estimated $1.6 billion for the Drinking Water SRF and $1.4 billion for the Clean Water SRF. USDA's Rural Utilities Service administers the Water and Waste Disposal program, which provides subsidized loans and grants for both drinking water and wastewater infrastructure projects in small communities. In fiscal year 2015, USDA made an estimated $1.7 billion available through the Water and Waste Disposal program. Small water utilities generally rely more on federal assistance to fund repairs and replacements to infrastructure than large water utilities, according to a 2012 Congressional Research Service report.[9]

In a March 2004 report, we found that water utilities may benefit from implementing asset management practices to better identify and manage their infrastructure needs.[10] Asset management is a widely recognized tool used across a variety of sectors to manage physical assets, such as highways, machinery, and buildings; in the case of water and wastewater infrastructure, key assets are pipelines, tanks, pumps, and other facilities. According to international asset management standards, asset management is a framework for providing the best level of service to customers at the lowest appropriate cost.[11] For example, in our March 2004 report, we found that, by implementing asset management, a large utility in California found that increasing preventive maintenance on its tanks would be more cost-effective than replacing them, saving the utility approximately $12 million. Federal law does not require water utilities to implement asset management, but federal law does require implementation of asset management in at least one other area. Specifically, federal law requires states to develop and implement an asset management plan for pavement and bridge assets that are part of the National Highway System.[12] Asset management is used by some states to manage real property and highways, by private organizations such as electric companies, and internationally in many sectors. Appendix I provides information about requirements and incentives for asset management.

In our March 2004 report, we found, among other things, that EPA had no central repository to facilitate information sharing on asset management within and across its drinking water and wastewater programs, which would help avoid duplication of effort. As a result, we recommended that EPA better coordinate its initiatives to promote asset management, explore asset management tools developed by other federal agencies, strengthen efforts to educate water utilities on how implementing asset management can help them comply with regulations, and establish a repository of information about asset management on the agency's website. EPA agreed with our recommendations and implemented them by, among other things, holding workshops and coordinating initiatives to provide asset management information. In a previous report, issued in August 2002, we surveyed water utilities serving populations larger than 10,000 on how these water utilities manage existing assets and plan for future improvements. We found that more than a quarter of the large water utilities we surveyed did not have asset management plans as recommended by water utility associations but had identified future infrastructure needs.[13] We did not survey small water utilities, and we did not make recommendations in that report. (See the "Related GAO Products" list for GAO's most recent reports related to asset management.)

In 2011, EPA and USDA signed a memorandum of agreement that stated that small water utilities face unique challenges in providing affordable drinking water and wastewater services. It noted that these systems face declining populations, lack financial resources and have not engaged in long-term planning activities, among other challenges.[14] By signing this memorandum of agreement, EPA and USDA agreed to collaborate in promoting ways that small water utilities can better manage their infrastructure needs and emphasized the use of asset management to ensure long-term technical, managerial, and financial capacity.[15]

In this context, you asked us to review water utilities' use of asset management to identify and fund capital improvements in response to rising infrastructure costs. Our objectives were to examine (1) what is known about the use of asset management among the nation's water utilities—particularly small water utilities—including benefits and challenges, if any, for water utilities implementing asset management and (2) steps, if any, that EPA and USDA are taking to help small water utilities implement asset management.

To examine what is known about water utilities' use of asset management, we interviewed EPA and USDA staff in the agencies' headquarters about the data they collect from water utilities, including data EPA collects in national assessments of water utilities' infrastructure needs that it conducts every 4 years. In addition, we interviewed state officials responsible for managing EPA's Drinking Water and Clean Water SRF programs as well as officials in USDA's state offices in a nonprobability sample of 10 states. To select these states, we used EPA's most recent needs assessment data (i.e., 2011 for drinking water infrastructure and 2008 for wastewater infrastructure)[16] to identify the state in each of EPA's 10 regions with the highest percentage of small water utility needs. These states were: Arizona, Arkansas, Delaware, Idaho, Iowa, Minnesota, Mississippi, New York, Vermont, and Wyoming. We reviewed relevant documentation and information from EPA staff responsible for maintaining the data to assess the reliability of both data sets and determined that they were sufficiently reliable for the purpose of selecting states and utilities for our review. Because our sample of states and utilities was nongeneralizable, responses from the officials we interviewed cannot be generalized to other states and their water utilities, but they illustrate some of the uses of asset management practices among small water utilities in states with the greatest infrastructure needs. We also interviewed representatives of national water and wastewater associations to identify potential sources of data on asset management. From these interviews, we identified one report and one on-going study. The report by McGraw-Hill Construction was published in 2013 and surveyed 451 persons representing water utilities on their use of asset management.[17] We reviewed the scope and methodology of this report and determined that it was sufficient for our purpose of describing qualitative information to corroborate information we obtained from our interviews about large utilities' use of asset management because the study authors included asset management practices that were similar to those identified by EPA. In 2015, the American Water Works Association surveyed officials with 545 water utilities about their use of asset management practices for the purposes of identifying additional resources the organization could provide to better assist water utilities wanting to start or advance an asset management program. The results of the survey were published in a report in December 2015.[18]

To examine the extent of asset management use in our sample of 10 states, we used EPA's 2008 *Asset Management: A Best Practices Guide*—a federally-developed asset management framework for water utilities.[19] The components and practices of the framework formed the basis of our interview questions. We used the practices in the framework to interview officials in a nongeneralizable, random sample of small utilities in the sample of 10 states. To select small water utilities in these 10 states, we used EPA's Safe Drinking Water Information System and a database of water utilities included in EPA's 2008 *Clean Watersheds Needs Survey* to randomly generate a nongeneralizable sample of two drinking water and two wastewater utilities in each of these states for a total of 40 small water utilities. Of these 40 small water utilities, we conducted interviews with officials from at least one drinking water utility and one wastewater utility in 9 of the 10 states, for a total of 25 water utility interviews.[20] The water utilities we contacted in Mississippi declined to participate or did not respond to our interview requests. We also visited Maine and New Mexico to conduct more in-depth interviews on water utilities' use of asset management. We selected these states based on recommendations from EPA and USDA officials, representatives of national water and wastewater associations, and technical assistance providers.[21] On these visits, we met with state SRF program and USDA officials, representatives of large and small water utilities, and technical assistance providers. To examine any benefits and challenges for small water utilities in implementing asset management, we analyzed the content of our interviews with state SRF program and USDA officials in the 10 selected states and in-person visits to Maine and New Mexico.

To examine steps EPA and USDA have taken to help small water utilities implement asset management, we reviewed EPA and USDA guidance, reports, training materials, and software tools and interviewed EPA and USDA officials. In addition, we interviewed technical assistance providers funded by EPA and USDA to conduct training sessions that included asset management. These technical assistance providers were the EPA-funded Environmental Finance Centers with the University of New Mexico, Wichita State University, Cleveland State University, and University of North Carolina at Chapel Hill and the national offices and local affiliates in Maine and New Mexico of the Rural Community Assistance Partnership and National Rural Water Association. We compared the information we collected about the steps EPA and USDA have taken with key practices related to federal agencies' training efforts and collection and dissemination of information that we identified in previous reports.[22] Appendix II contains a more detailed description of our objectives, scope, and methodology.

We conducted this performance audit from January 2015 to January 2016 in accordance with generally accepted government auditing standards. Those standards require that we plan and perform the audit to obtain sufficient, appropriate evidence to provide a reasonable basis for our findings and conclusions based on our audit objectives. We believe that the evidence obtained provides a reasonable basis for our findings and conclusions based on our audit objectives.

BACKGROUND

This section provides background information on (1) asset management for water utilities, (2) federal funding for asset management, (3) water utilities' structures, and (4) EPA's infrastructure needs assessments.

Asset Management for Water Utilities

To assist water utilities in adopting asset management, in 2003, EPA developed an asset management framework for water utilities.[23] In 2008, EPA incorporated this framework into a best practices guide for water utilities based on similar frameworks used by water utilities in Australia and New Zealand.[24] EPA's asset management framework instructs water utilities to (1) assess the current state of their assets, (2) determine the level of service they need to provide to customers, (3) identify those assets that are most critical to their operations, (4) incorporate life-cycle costs, and (5) develop a strategy for the long-term funding of the repair and replacement of their assets. As shown in figure 1, EPA's 2008 best practices guide describes the five components in EPA's asset management framework, which are characterized by a range of practices. According to EPA's best practices guide, these practices can be implemented at varying levels of sophistication, depending on the size and needs of the utility. For example, a small water utility with few assets can document its inventory of assets on paper, although a large water utility with many assets may use a software program.

Asset management for water utilities

In its 2008 *Asset Management: A Best Practices Guide*, the Environmental Protection Agency (EPA) defines asset management as a framework for maintaining a desired level of service for what a water utility wants its assets to provide at the lowest life-cycle cost. Lowest life-cycle cost refers to the best appropriate cost for rehabilitating, repairing, or replacing an asset. Examples of a water utility's assets are: pipes, pumps, valves, and tanks. The asset management framework consists of five core components, which comprise a range of practices. The five components with some of EPA's examples of practices are:

#1 Current state of assets

- preparing an asset inventory and map of the location of assets
- assessing the condition of assets
- determining asset values and replacement costs

#2 Level of service

- analyzing current and anticipated customer demand
- writing and communicating a public level of service with performance targets

#3 Critical assets

- listing assets according to how critical they are to operations
- conducting a failure analysis
- determining probability of failure, failure risk, and consequences
- reviewing and updating vulnerability assessments

#4 Minimum life cycle cost

- moving from reactive to predictive maintenance
- determining the costs and benefits of rehabilitation versus replacement
- reviewing life cycle costs for critical assets
- deploying resources based on asset conditions
- developing specific response plans based on causes of failure

#5 Long-term funding plan

- revising the rate structure
- funding a dedicated reserve from current revenues

Figure 1. EPA's Asset Management Framework and Component Practices.

Together, according to EPA's 2008 best practices guide, these practices make up a water utility's asset management program and are to be documented in the water utility's asset management plan. The asset management plan serves as a written record that the water utility can use much like a budget or strategic planning document to communicate plans, progress, and future goals and also communicate user rate adjustments and recommended infrastructure investments. According to an EPA fact sheet on building an asset management team,[25] asset management requires water utility staff who can promote and articulate the benefits of asset management.[26] The fact sheet further states that a successful asset management program requires resources, including time and money, to implement, as well as the support of political leaders who have the authority and willingness to commit public resources and personnel.

We and others have cited examples of cost savings resulting from asset management. In our March 2004 report,[27] in addition to the water utility in California that saved $12 million, we found that a water utility in Massachusetts used asset management and saved $20,000 in oil purchase and disposal costs for its pumps and decreased the hours spent on preventive maintenance by 25 percent from the hours recommended by the equipment manufacturer. In addition, a 2007 study of asset management by the U.S. Conference of Mayors also found that public water utilities in cities had experienced savings in capital costs and operations and maintenance as a result of asset management.[28] Further, a 2008 EPA fact sheet about asset management for local officials stated that implementing asset management may require some up-front costs but could result in cost savings for water utilities.[29]

In their 2011 Memorandum of Agreement, EPA and USDA agreed to collaborate in promoting ways that small water utilities could better manage their infrastructure needs and highlighted the use of asset management to ensure long-term technical, managerial, and financial capacity. The agencies also agreed to coordinate agency activities and financial assistance in areas that would increase the technical, managerial, and financial capacity for small water utilities. The memorandum stated that EPA and USDA would encourage communities to implement system-wide planning, including asset management, and that the two agencies would share and distribute resources to water utilities and provide training and information. In this same memorandum, EPA and USDA stated that both agencies supported increasing the technical, managerial, and financial capacity of water utilities nationwide.

Federal Funding for Asset Management

EPA and USDA funding for asset management activities falls under various larger programmatic budget categories. EPA funds asset management in the following three categories: (1) grants to provide training and technical assistance to water utilities to improve financial and managerial capacity; (2) grants to selected public, private universities or colleges, and nonprofit organizations to provide technical assistance to communities on a range of EPA priorities, including improving financial capacity; and (3) drinking water SRF grants to states, a portion of which may be used for increasing water utilities' technical, managerial, and financial capacity. USDA primarily funds asset management activities through two programs: (1) the Water & Waste Disposal Technical Assistance & Training Grants program, which provides grants to nonprofit organizations in the 50 states for managerial technical assistance, and (2) the Circuit Rider program, which provides training and technical assistance through contracted staff called circuit riders,[30] in each of the 50 states to provide technical assistance to small water utilities on day-to-day operational, managerial, and financial issues. Appendix III provides more details about EPA and USDA funding for asset management.

Water Utilities' Structures

Small communities share some common characteristics in how they manage (govern and staff) their water utilities, according to EPA's 2011 report on the characteristics of small water utilities.[31] The report, and EPA, have made several observations about small water utilities and the small communities they serve. Namely, publicly-owned water utilities are typically municipalities, townships, counties, or other public entities. These entities can be governed by boards, mayors, managers, or city or town councils.[32] Privately-owned water utilities are typically governed by corporate entities, homeowner associations, or sole proprietors. For both publicly- and privately-owned water utilities, the governing bodies are responsible for ensuring the water utility complies with state and federal laws and regulations; setting and approving annual budgets; hiring staff; and in many cases, setting and adjusting the rates that users pay. EPA's 2011 report also states that small water utilities are typically staffed with an operator (or superintendent), managers, and administrative staff that may work part-time. In some cases, a publicly-owned water utility may hire a private company to operate and maintain its facility.

EPA Infrastructure Needs Assessments

EPA estimates the nation's drinking water and wastewater utilities' capital infrastructure needs over the next 20 years by administering to states two needs assessment surveys: the *Drinking Water Infrastructure Needs Survey and Assessment* and *Clean Watersheds Needs Survey* every 4 years.[33] In completing the questionnaire for the drinking water needs assessment survey, utilities report infrastructure needs to EPA and for the clean water needs assessment survey states report these infrastructure needs to EPA. EPA then uses the data from the drinking water needs assessment survey to determine each state's grant allocation for its Drinking Water SRF program. According to EPA officials, the agency does not use the clean water needs assessment survey to determine each state's allocation of Clean Water SRF program funds, but it reports the data to Congress and the public.

EPA works with states and the Office of Management and Budget (OMB) to produce the surveys.[34] The questionnaires for both needs assessment surveys ask about water utilities' infrastructure needs, including those assets that are in need of replacement or rehabilitation. EPA officials said that they accept certain documents as support for the states' cost information, including SRF loan applications, capital improvement plans,[35] and asset management plans. To support their reported infrastructure needs, some water utilities submitted documentation that showed the use of asset management practices, according to the results of the 2011 drinking water needs assessment survey, and other utilities' supporting documentation illustrated continuing gaps in knowledge about the condition and remaining useful life of their infrastructure.[36] The results of the 2008 clean water needs assessment survey also highlighted water utilities' use of asset management and featured examples of states that used asset management practices to determine the costs of projects submitted to EPA.[37]

Small Water Utilities in Selected States Are Implementing Some Asset Management Practices, and EPA, USDA, and State Officials Have Identified Benefits and Challenges

The small water utilities we interviewed in our sample of 10 states are implementing some asset management practices, and the state SRF officials we interviewed in these states said that large water utilities are more likely to implement asset management practices than small water utilities. EPA, state SRF, and USDA officials in our review identified benefits that could result from water utilities' use of asset management practices, as well as challenges water utilities face in implementing them.

Small Water Utilities in Selected States Are Implementing Some Asset Management Practices

Officials we interviewed from small water utilities in the selected states said that they are implementing some asset management practices, and state SRF program officials in these selected states indicated that large utilities are generally more likely to implement asset management. Officials we interviewed at the 25 small water utilities we selected for our review generally told us they were implementing some of the asset management practices EPA identified in its 2008 asset management best practices guide, but we found differences in the extent to which these small water utility officials were implementing these practices. We discuss what we found using EPA's framework, which consists of the five components and a written asset management plan.

- **Current state of assets**. EPA's 2008 best practices guide states that water utilities, in assessing the current state of their assets, should know what assets they own, what condition they are in, and their remaining useful life—that is, how much longer the water utility expects their assets to last. EPA recommends that water utilities (1) compile this information into an asset inventory that lists each asset's age, condition, service history, and remaining useful life and (2) develop maps that identify the location of these assets. Officials we interviewed from 8 of 25 of the small water utilities we reviewed in the selected states told us they had an inventory listing all of their assets, and 19 of 25 told us they had an inventory that listed at least some of their assets. Of the 8 small water utilities that had complete asset inventories, 2 of 8 included information on each asset's physical condition, and 3 of 8 included an estimate of each asset's remaining useful life, according to the officials we interviewed with these utilities. These officials described various types of inventories, ranging from a list of assets included on insurance documents to a software program that included information about the assets' age, condition, service history, and remaining useful life. Officials at almost all (23 of 25) of the small water utilities in the selected states told us they had maps that identify the location of at least some of their assets. These officials described a range in the types of maps used by their water utilities, including one official who described using maps dating back to the 1980s, but others described using maps generated with a geographic information system (GIS) to locate a water utility's assets.[38]

- **Level of service.** EPA's 2008 best practices guide states that water utility operators need to know the level of service they will provide—that is, (1) what customers and stakeholders demand, (2) what regulators require, and (3) what operators need to know about the actual performance and capabilities of the water utility itself. According to EPA's 2008 guide, water utilities should also set performance goals related to these three facets of service. Officials at (1) 11 of 25 of the small water utilities in the selected states told us they had performance goals related to customer demand, (2) 19 of 25 said they had performance goals related to meeting EPA and state regulations, and (3) 17 of 25 said they had performance goals related to the actual performance of the system. For example, at a small water utility, an official responsible for managing the community's water utilities described setting goals to control the loss of treated drinking water from leaky distribution pipes and the loss of untreated wastewater through leaks in the sewer system. According to EPA's website, leaks in sewer systems can result in sewage overflows, increasing the quantity of water requiring treatment—which, in turn, can increase a wastewater utility's costs and present public health and environmental risks. The official told us that the water utility compares the amount of water that the drinking water utility produces to the amount of water used by customers, as indicated by their individual water meters, to ensure that no more than 20 percent of the water is lost through leaky distribution pipes. Additionally, this official described comparing drinking water production to flows from the town's sanitary sewer and wastewater treatment plant to ensure that these overall flows are neither too high nor too low and no less than 90 percent of the community's drinking water eventually makes it to the wastewater utility for treatment. Other officials we interviewed in the selected states described a range of goals, some of which did not relate to asset management. For example, some officials told us that their goals related to customer demand were to simply keep the water utility operating or to meet peak customer demand.

- **Critical assets**. EPA's 2008 best practices guide states that water utilities need to know which assets are the most critical to sustaining the water utility's operations, their risk of failing, and the consequences if they do fail. Officials from 18 of 25 of the small water utilities in our selected states told us they had identified their water utility's critical assets, but officials from 11 of 25 of those utilities said they have assessed the probability of failure for every critical asset. Officials we interviewed in 15 of 25 of the utilities in the selected states told us that, generally, the likelihood and consequences of failure for assets informed their decisions about which infrastructure projects to fund. Officials we interviewed in the selected states described taking a range of approaches to identify and assess their critical assets. For example, an official from one small water utility described the process of identifying and assigning a score (i.e., minor, major, or catastrophic) to each critical asset based on the impact that asset's failure would have on the environment and customer needs. An official with another small water utility described having enough experience with the water utility to keep mental notes about which assets were critical to the water utility's operations, and another official described using an online, computer-based system to operate critical assets remotely to monitor the probability of failure.
- **Minimum life-cycle cost**. EPA's 2008 best practices guide states that asset management enables a water utility to determine the minimum life-cycle cost—that is, the lowest cost options for providing the highest level of service over the lifetime of an asset. According to the guide, water utilities can achieve this by scheduling operations and maintenance based on the condition of assets; knowing the costs to repair, rehabilitate, and replace assets; and having specific response

plans in case assets fail. Officials from 19 of the 25 small water utilities we reviewed in the selected states told us they conduct regular maintenance, but officials from 9 of 25 said they knew the cost of rehabilitation versus the cost of replacement for all water utility's assets. For example, one official said that the water utility had not determined the costs of rehabilitation versus replacement because the assets were too old to be considered for rehabilitation. Additionally, officials from 15 of the 25 small water utilities in our selected states had written plans that describe their water utility's response in the case of asset failure. Concerning written plans to address asset failure, one official described a plan outlining discrete protocols the water utility should take to address asset failures or emergencies, but another official described a list of individuals or repair companies the water utility should notify when an asset fails.

- **Long-term funding plan.** EPA's 2008 best practices guide states that asset management activities related to developing long-term funding plans involve determining whether the water utility has enough funding to maintain its assets based on the required level of service (i.e., customer demands, regulatory requirements, and the capability of the utility's assets) and whether the user rates are sufficient for the water utility's long-term needs. EPA's 2011 report on the characteristics of small water utilities described communities and water utilities as generally separating funds by routine operations and maintenance and capital improvements and that they may also have an emergency fund or reserve fund earmarked for a specific purpose.[39] Officials at 19 of the 25 small water utilities in the selected states told us they had established a reserve fund to cover the cost of short-lived assets,[40] but officials at 11 of the 25 small water utilities told us they had enough funds to cover their water utility's long-term capital infrastructure needs. For example, an official from one small water utility described using a two-tiered rate structure consisting of a monthly water usage rate and a depreciation fee. This official said that the water utility uses the monthly rate to cover operations and maintenance and short-term capital infrastructure costs and sets the depreciation fee aside to fund long-term capital infrastructure costs. Other officials from small utilities described a range of approaches to planning for the long-term. For example, some small water utility officials told us that the water utility established separate reserves for short- and long-term capital investment needs, but an official from one small water utility described establishing a general surplus account into which the water utility put any surplus funds available at the end of the year for repairs and replacement.

- **Asset management programs and plans.** According to EPA's 2008 best practices guide, asset management is implemented through an asset management program and typically includes a written asset management plan. Officials at the small water utilities we interviewed said that they are implementing asset management practices as a routine course of business rather than a concerted effort to implement a formal asset management program or plan. Therefore, officials at 5 of the 25 small water utilities in the selected states said that they had a written asset management plan.

The small utilities in our selected states were implementing some asset management practices, but officials we interviewed with 9 of the 10 state SRF programs in our selected states told us that, generally, the large water utilities in their states were more likely than small water utilities to implement asset management.[41] Similarly, a 2013 market research study by McGraw-Hill Construction found that larger water utilities were more frequently implementing asset management practices than smaller water utilities.[42] Officials from the large utility we interviewed in Maine and the large utility we interviewed in New Mexico said they were implementing what they considered as comprehensive asset management practices, that is, practices as outlined in all five components of EPA's framework. For example, officials from a large water utility in Maine said that it had a performance goal for the district's fire hydrants related to the level of service provided to customers—that is, all fire hydrants would be in working order and would not spend more than 3 days out of service. These officials said that the inspections of the fire hydrants were electronically tied to the asset management software, which allows the water utility's managers to monitor the status of the inspections and track progress related to the performance goal.

EPA, USDA, and State Officials Identified Benefits Resulting from Water Utilities' Use of Asset Management Practices and Challenges in Implementing Such Practices

EPA and USDA headquarters officials and state SRF and USDA officials cited benefits for both water utilities and federal agencies resulting from water utilities' use of asset management practices. They also cited challenges for water utilities—particularly small water utilities—in implementing asset management practices, particularly the costs of implementing such practices.

Benefits

EPA and USDA headquarters officials and state SRF and USDA officials cited benefits for water utilities that implement asset management, including: (1) cost savings for water utilities that prolong the useful life of their assets and avoid costly emergency repairs, (2) more efficient, focused long-term planning of management and operations, and (3) improved financial health for water utilities. They also cited benefits for federal agencies.

Cost Savings

EPA headquarters and state SRF and USDA officials told us that water utilities implementing asset management can experience cost savings by prolonging the useful life of the assets they already own through preventive maintenance, including pipe lining and repair, and deferring replacement costs. EPA's guidance states that preventive maintenance can help water utilities avoid unnecessary additional costs. Officials in our review provided the following examples:

- An official from one small water utility in Maine told us that the process of creating an asset inventory helped the water utility identify assets they did not know they owned and therefore had not maintained. This official also told us that the utility's use of asset management helped utility staff assess the condition of the utility's assets and implement a regular preventive maintenance schedule to maintain those assets. According to the official, this helped the utility avoid larger replacement costs, but he could not estimate the amount of savings.
- Another official with a small water utility in Idaho told us that the utility used asset management to plan the maintenance and repair of its drinking water reservoir and fire hydrants, which extended their useful life and resulted in cost savings.
- Another official with a small water utility in Maine told us that the utility assessed the condition of its sewer lines and realized that hydrogen sulfide —a result of the type of materials used to construct the pipes— had built up and put several lines at risk of collapsing.

The official said the utility spent $12,000 to remove the hydrogen sulfide and prevented the collapse.

More Efficient and Focused Long-Term Planning of Operations and Management

State SRF and USDA officials said that water utilities implementing asset management can plan more efficiently for the long-term, such as planning for capital investments, identifying changes in infrastructure needed as a result of population change, hiring or succession planning, improved emergency planning; and making decisions about repairs and replacements. The officials we interviewed highlighted the following examples:

- Officials at a small water utility in New Mexico told us that having an asset management plan allowed the utility to prioritize its capital investment needs, identify the associated costs, and determine what resources it would need going forward.
- An official with a small water utility in Arkansas told us that the utility used its asset management plan to assess the effect of a new housing development on its drinking water and wastewater infrastructure over the next 5 years. As a result of this assessment, the water utility was able to set connections and new user fees to recover the costs of adding the housing development without increasing water utility rates for existing residents.
- An official with a small water utility in Maine told us that because it identifies assets, maintenance schedules, and replacement schedules, creating an asset management plan was the best way for water utilities to transfer decades of knowledge retiring operators and maintenance staff had about the system. The official said that ensuring the continuity of operations and service to the community after employees retire provides some long-term planning.

Improved Financial Health

State SRF and USDA officials told us that water utilities implementing asset management can improve their financial health. EPA headquarters officials said that asset management can help water utilities better budget for capital investments and justify increases in user rates. Asset management also enables water utilities to better account for the value of their capital assets and asset depreciation, which can improve financial transparency and help the utilities with the documentation needed for financial audits. The officials we interviewed highlighted the following examples:

- An official at a small water utility in Maine told us that the utility used its asset management plan to determine its financial needs, calculate a new user rate to meet these needs, and successfully justify raising rates to the water utility board and its customers.
- Another official at a small water utility in New Mexico told us that the water utility uses its asset management program to track its finances, including the depreciation of its assets—information that is typically reviewed as part of its financial audit.

Benefits to Federal Agencies

According to state SRF and USDA officials we interviewed, the federal agencies with programs that provide loans and grants to small water utilities to help fund capital infrastructure can also benefit from water utilities' use of asset management, as follows:

- State SRF and USDA officials said that a benefit of asset management, for lenders, is knowing that federal funds would be better targeted for infrastructure projects that address a community's greatest needs and knowing that the federal funds are paying for a project that the community could not afford on its own.
- EPA officials stated that increased use of asset management by small water utilities would improve the utilities' assessments of their capital needs, thereby improving the quality of the data collected for EPA's needs assessments. In addition, EPA officials we interviewed said that water utilities' use of asset management can result in more accurate information about infrastructure needs, such as costs, and better management of the funds spent on infrastructure repairs and replacement.[43]

Challenges

In addition to benefits, the state SRF and USDA officials we interviewed generally identified the following key challenges small water utilities face in implementing asset management:

- **Costs**. According to EPA's 2008 guidance on asset management for local officials, implementing an asset management program may include start-up costs. For example, SRF officials in one state told us that start-up costs are the largest costs for water utilities, often challenged with limited resources, in implementing asset management. According to state SRF and USDA officials, start-up costs can include (1) purchasing asset management tools, such as software or creating GIS maps, or (2) hiring an engineer or consultant to create an asset management program or plan on water utilities' behalf. For example, officials with two separate small water utilities in New Mexico told us that they spent $34,000 and $50,000, respectively, to hire a company to create GIS maps of the water utility's assets and officials with another small water utility in New Mexico told us that they paid an engineer $12,000 to develop an asset management plan.
- **Funding**. State SRF and USDA officials we interviewed said that small water utilities have difficulty obtaining funds or anticipate they will have difficulty obtaining funds to cover the start-up and maintenance costs associated with asset management. In describing challenges with funding asset management, for example, officials with a small water utility in New Mexico told us that the utility did not have the funds to pay an engineering firm to develop the needed additional GIS maps with the locations of their assets and would have to apply to a state infrastructure grant program for an additional $50,000.
- **Human resources**. According to the state SRF and USDA officials we interviewed, small water utilities do not have human resources to dedicate to asset management. For example, officials with a small water utility we visited in Maine said that, at the time of our review, the Maine Department of Transportation was completing a major road project in the state that affected the buried pipes for multiple communities, including the one in which this utility operated. These utility officials said that work on these pipes in addition to the routine day-to-day responsibilities of operating the utility left the small staff little time to work on asset management. Similarly, an official with another small water utility in Maine told us that one staff person was assigned to develop an inventory of the water utility's assets, and that finding the time was the greatest constraint to completing the

inventory, coordinating with operations and maintenance staff, and implementing additional asset management practices.

- **Information**. Acquiring information about how to start or maintain an asset management program was another challenge for small utilities that state SRF and USDA officials cited. For example, officials with a small water utility in New Mexico said that the town leadership was unaware of asset management prior to applying for an infrastructure loan through a state program. As a result, it took some time for the water utility operator and the utility's board to understand the asset management concept and implement the activities required as part of the state's loan program.
- **Political support.** According to some of the state SRF and USDA officials we interviewed, small water utilities are challenged with garnering and maintaining the political support of elected officials and the local community to begin or maintain an asset management program or increase user rates or expend funds on repairs as a result of implementing an asset management program. For example, an official with a small water utility said that the town's council was supportive of the recommendations the utility operator made regarding the likelihood of failure for assets and the need to address those assets before they failed. However, the town council did not always implement the recommendations because, among other things, they said they wanted to avoid having to raise user rates to cover the costs. An official with another water utility said that the water utility would benefit from raising user rates incrementally each year, but that elected officials do not want to raise rates, even minimally, because the community would not support such raises.

EPA and USDA Are Taking Steps to Help Small Utilities Implement Asset Management but Do Not Collect Information on Asset Management to Track Its Use or Compile Information on Costs and Benefits

EPA and USDA are taking steps to help small utilities implement asset management and address identified challenges that water utilities face. EPA and USDA recognize the benefits to water utilities and their loan programs and the need for water utilities, particularly small water utilities, to increase their use of asset management, but the agencies do not collect information on asset management that would enable them to track their efforts or compile information on costs and benefits that could be used to encourage wider use.

EPA and USDA Are Taking Steps to Help Small Water Utilities Implement Asset Management

EPA and USDA officials told us that they would like for as many water utilities as possible to increase their managerial and financial capacity, including the implementation of asset management. The officials said they are aware that small water utilities face challenges in implementing asset management and are taking steps to help them. To help small water utilities implement asset management, EPA and USDA provide funding for the development of asset management plans, free or low-cost tools such as software to develop asset management programs and plans, classroom training, and one-on-one technical assistance or coaching.

Funding

Both agencies provide funding for the development of asset management plans, helping to address the challenges of costs and funding. EPA provides funds that can be used for the development of asset management plans through grants to state drinking water SRF programs. According to state SRF officials in some of our selected states, these funds can help water utilities address challenges in finding the funds to cover the start-up costs of asset management activities. For example, officials with the Maine Drinking Water SRF program told us that the state SRF program uses its drinking water SRF funds[44] to pay up to 75 percent of the cost of developing an asset management plan for water utilities serving populations of fewer than 3,300 people and up to 50 percent for water utilities serving populations of more than 3,300 people. According to these officials, about 15 water utilities have applied for this funding between 2013 and 2015. State officials with the Delaware Drinking Water and Clean Water SRF programs, for example, told us that they recently started a new program providing grants to water utilities for funding activities leading to the development of an asset management plan. As of June 2015, the state had provided a grant ranging from $60,000 to $100,000 to each of the 4 publicly-owned water utilities that participated in the program.

State SRF officials in some of the 10 states in our review told us that in meeting state requirements for SRF loans,[45] small water utilities in their states were engaging in some asset management practices.[46] State officials we interviewed provided examples, such as requiring (1) a report of the inventory and condition of the utility (or preliminary engineering report)[47] to show technical capacity; (2) a community to raise user rates to pay back the loan; or, (3) a community to set up a reserve fund to pay for short-lived assets. Officials we interviewed at 10 of the 25 small water utilities in the selected states said that they currently had an SRF loan.

USDA officials in the agency's headquarters and all 10 state offices we interviewed also told us that, as a result of their loan requirements,[48] small water utilities with USDA loans were engaging in some asset management practices. USDA headquarters officials told us that asset management is incorporated throughout their loan conditions.

Specifically, USDA state officials said that they consider the following loan conditions to equate to asset management practices: requiring (1) a review of financial audits, (2) a preliminary engineering report, (3) a community to create a reserve to fund debt payments and cover the repair and replacement of short-lived assets, (4) development of an operations and maintenance manual, and (5) the restructuring of user rates to cover the cost of the loan and repair and replacement of short-lived assets. According to some USDA headquarters and state officials, USDA's state offices also conduct periodic (every 3 years) inspections of the condition of the facilities they fund once they are built. Officials we interviewed at 6 of the 25 small water utilities in the selected states said that they currently had a USDA loan.[49]

USDA officials said that their use of the preliminary engineering report is the key way in which the agency introduces its loan applicants to asset management. In 2013, USDA, in conjunction with EPA and other federal agencies and states, issued a preliminary engineering report template, a planning document that, in general, includes an inventory of the category of assets and assessment of the assets in the entire facility (e.g., the assets involved in the project being funded, a map of the assets in the water utility); information about the need for the project (including most critical aging infrastructure and future growth needs); and the costs for the repair, rehabilitation, and replacement of some assets. USDA regulations require loan applicants to submit a preliminary engineering report, and encourage applicants to consult agency guidelines in preparing the report.[50] In a 2013 bulletin to state officials, USDA encourages its state offices to use the preliminary engineering report template. EPA does not require SRF loan applicants to submit a preliminary engineering report, but like USDA, it encourages its use; specifically, it encourages state SRF programs to require its use.[51] EPA officials told us that as of October 2015, 10 state SRF programs had adopted the preliminary engineering report template and 10 other state SRF programs had adopted it and modified it by including additional requirements.

Free or Low-Cost Tools

Both agencies provide free or low-cost tools for developing asset management programs and plans to help address the challenges of cost and providing information. EPA provides a free asset management software program, and both EPA and USDA provide free tools such as guidebooks, case studies, and other written materials for small water utilities on the agencies' websites. EPA's free software program, Check Up Program for Small Systems (CUPSS), allows water utilities to develop asset management programs and plans. EPA officials told us that the original development of CUPSS was funded with SRF funds. Users of CUPSS can enter data into the system to develop an inventory of assets, record information to track the scheduling of maintenance tasks, and produce a written asset management plan. EPA officials told us that, with CUPSS, water utility managers can produce a report specifically communicating the condition of the water utility's assets to elected officials. EPA also provides free training on how to use CUPSS. The availability of CUPSS also allows utilities to avoid some of the costs they would incur if they were to hire a professional engineering firm to do the same work. For example, an official with a small water utility we interviewed said that he did not incur any monetary costs to implement asset management because he used CUPSS to develop his asset management plan and program.

Classroom Training

EPA, USDA, and state SRF programs provide classroom training on asset management to help provide information to operators and other staff about how to implement asset management.

- EPA's Environmental Finance Center (EFC) [52] at the University of North Carolina at Chapel Hill leads the Smart Management for Small Water Systems project that provides 1-day workshops for operators of water utilities on various aspects of managing a water utility. According to an official with the EFC at the University of North

Carolina at Chapel Hill, the EFC partners with other EFCs (including the University of New Mexico and Wichita State University) to conduct the workshops, which include a discussion of asset management. As stated on the website for the EFC at the University of North Carolina at Chapel Hill, from 2012 to 2014, these EFCs held more than 100 workshops, with 2,000 participants, in all 50 states and four U.S. territories.

- According to USDA headquarters officials, operators of small water utilities and their elected officials participate in financial and managerial training courses provided by organizations, such as the National Rural Water Association,[53] Rural Community Assistance Partnership,[54] and others. According to these officials, these training sessions can include asset management. An official with the National Rural Water Association told us that the organization's training sessions generally include a component of asset management. Officials with the Rural Community Assistance Partnership told us that their organization provides workshops specifically on implementing asset management, including workshops for elected officials.
- EPA and USDA officials told us that the agencies' key collaborative effort is a workshop on water utility management, with the goal of helping to increase water utilities' managerial and financial capacity. The workshop, based on a 2013 EPA and USDA document entitled *Rural and Small Systems Guidebook to Sustainable Utility Management*,[55] describes 10 steps in effectively managing a water utility. EPA and USDA officials said that asset management is discussed as part of 1 of the 10 steps. EPA and USDA's guidebook defines the steps, describes challenges water utilities may face related to the steps and the effects of those challenges, and describes the types of actions taken by high-performing water utilities to address the challenges. The workshop materials focus primarily on the logistics of implementing a workshop. EPA and USDA train technical assistance providers to conduct the workshop for water utilities. Both agencies also provide free materials for the workshop on their websites. A 2015 EPA and USDA report[56] stated that the agencies had trained1,600 persons in workshops across the United States since 2013. EPA agreed with our estimate that the two agencies, together, train about 250 water utilities per year. USDA officials told us that for fiscal year 2015, the agency provided a grant to a technical assistance provider to

provide two workshops in each of the 50 states. EPA funds this effort through its Small Systems Training and Technical Assistance Grants program, and USDA funds this effort through its Water & Waste Disposal Technical Assistance & Training Grants program.

- State SRF programs also use some of their federal funds to provide classroom training, for small water utilities, on a variety of topics related to building small water utilities' managerial and financial capacity, including asset management. For example, according to Maine Drinking Water SRF officials, Maine's Drinking Water SRF program provided four $25,000 grants to a local technical assistance provider to train operators of small water utilities and their elected officials on asset management.

This classroom training provides water utilities with education and information about asset management and how to implement an asset management program.

One-on-One Technical Assistance or Coaching

Both agencies also provide one-on-one technical assistance or coaching on asset management, which helps address the challenges of costs, funding, and providing information. EPA, USDA, and state SRF programs work with many of the same organizations to provide technical assistance services in their states. EPA also reaches small water utilities through its EFCs. The two primary organizations with whom EPA and USDA work to reach small water utilities are the National Rural Water Association and Rural Community Assistance Partnership. EPA and USDA officials have said that their contracts for technical assistance with these two providers are not exclusively for asset management, but that technical assistance providers are trained to and frequently help water utilities implement asset management and develop asset management plans. In conjunction with its workshop on sustainable utility management, EPA and USDA officials told us that technical assistance providers, in their workshops, also conduct follow-up calls to workshop participants and, if necessary, provide one-on-one assistance. The availability of one-on-one technical assistance also allows water utilities to avoid some of the costs they would incur if they were to hire a professional engineering firm to do the same work. For example, an operator with a small water utility in Maine told us that the utility developed its asset management program and plan through CUPSS with the free help of an organization contracted with EPA and USDA to provide technical assistance. This operator said the water utility's asset management program would not have been developed without this technical assistance.

EPA and USDA Do Not Collect Information on Asset Management to Track Their Efforts or Compile Information on Costs and Benefits to Encourage Wider Use

EPA and USDA recognize the benefits of asset management to water utilities and their loan programs and the need for water utilities, particularly small water utilities, to increase their use of asset management. Both agencies—EPA since 2003 and USDA since 2011— have identified asset management as a tool that water utilities can use to increase their ability to address infrastructure needs. In their 2011 memorandum of agreement, EPA and USDA agreed to collaborate in promoting ways that small water utilities can better manage their infrastructure needs and highlighted the use of asset management to ensure long-term technical, managerial, and financial capacity. EPA and USDA agreed to coordinate agency activities and financial assistance in areas that would increase the technical, managerial, and financial capacity of small water utilities. The memorandum also stated that EPA and USDA would encourage communities to implement system-wide planning, including asset management, and that the two agencies would share and distribute resources to water utilities and provide training and information. EPA and USDA officials told us that they want their efforts to result in as many water utilities as possible increasing their managerial and financial capacity, including the use of asset management. However, even though EPA and USDA promote sustainable water infrastructure, and the agencies encourage water utilities to better manage their resources to address the long-term challenges posed by deteriorating infrastructure, limited funds, and declining populations, they do not—and are not required to—collect information on utilities' use of asset management. Specifically, they do not collect information that tracks the results of their training efforts on utilities' use of asset management practices or compile information on the benefits and costs of implementing asset management.

First, EPA and USDA do not collect information that tracks the results of the agencies' training efforts (e.g., whether participating utilities use asset management practices). EPA stated in a 2011 policy document describing its plans to promote sustainable water infrastructure that the agency has an interest in tracking the results of the agency's training. This is consistent with our January 2004 report on selected agencies' experiences and lessons learned in designing training and development programs and our March 2004 guide on assessing strategic training and development efforts for human capital,[57] in which we reported that evaluating training programs is key to ensuring that training is effective in contributing to the accomplishment of agency goals and objectives. It is also consistent with our September 2005 report on enhancing performance management, which states that information can be used to make decisions that affect future strategies, planning and budgeting, identifying priorities, and allocating resources.[58]

Both EPA and USDA collect some information from the water utilities that participate in classroom training or receive one-on-one technical assistance; however, the agencies do not collect information that may allow them to better measure the results of their efforts to assist utilities. EPA collects information on the number of utilities that have taken training per year and reports this as part of its major performance goals. For example, as stated above, since 2011, 250 small utilities per year have taken the EPA and USDA trainings (of the 68,000 utilities nationwide). In addition, both EPA and USDA collect feedback from water utilities on their experience in training. For example, the agencies collect feedback forms from those attending the training to determine how to improve it.

However, the information EPA collects does not show whether the water utilities that receive training from the agencies went on to incorporate asset management practices into their work processes or whether these water utilities have improved their managerial or financial capacity.

EPA and USDA officials said that information on water utilities' use and incorporation of asset management would help EPA and USDA understand how the training and technical assistance they provide are affecting utilities' use of asset management. EPA officials said that they would like to collect data on water utilities' incorporation of asset management and determine whether these water utilities have improved their managerial or financial capacity, but they do not have the resources to do so. An EPA official said that when the agency was considering a nationwide study in 2006, the agency wanted to study the incorporation, costs, and benefits, of water utilities' use of asset management. However, EPA found that the costs of such a data collection effort would be in the hundreds of thousands of dollars. In particular, according to an EPA official, in addition to the costs of collecting the data from water utilities, the agency would also face costs in submitting the required Information Collection Request proposal to OMB. USDA officials similarly stated that they would be interested in collecting data on how the water utilities that participate in their sustainability workshops incorporated asset management or other management practices into their work processes. However, a USDA official told us that the agency would have to explore whether it could do a study such as this.

Leveraging existing data collection methods to collect information, such as adding questions to existing information collection requests, could be a cost-effective option for EPA and USDA to obtain information on water utilities' use of asset management. In particular, EPA's drinking water and clean water infrastructure needs assessment surveys provide national data on water utilities' infrastructure repair, rehabilitation, and replacement needs. EPA officials said that EPA works in partnership with states to obtain the data, and it works with state officials to convey the importance of the surveys. This work includes discussing the types of questions in the surveys, including any additional policy areas the questions will cover. EPA officials told us that the agency can add questions to the needs assessment surveys and has done so in the past on such policy issues as climate change and energy efficiency; the officials said that few states have responded to the questions, however, because participation by the states and water utilities is voluntary, and states are not required to answer all of the questions in the surveys.

EPA officials said that the agency included questions about asset management in the clean water needs assessment survey in 2008 and 2012. Specifically, these officials said, the questions asked water utilities to identify the status of their implementation of asset management and related costs. According to these officials, the agency did not receive enough responses to analyze and include the data in the final reports. EPA officials representing the clean water needs assessment survey said that they would be open to more discussions with states about asset management given their increased awareness. An EPA official that works with the drinking water needs assessment survey told us that EPA has not systematically asked questions about asset management in the drinking water needs assessments survey because the agency has determined that its efforts are best focused on asking questions required to determine infrastructure needs and for which the agency is likely to receive a large response from the states. In addition, this official said that EPA encourages water utilities to implement asset management practices through its guidance to states about the types of information the agency will accept as support for responses to the survey. An EPA official said that the agency has not considered or determined what asset management questions it might ask on the drinking water needs assessment survey. By continuing to include questions on the clean water needs assessment and considering questions about water utilities' use of asset management to include in the drinking water needs assessment survey, EPA may have better assurance that it is collecting information in a cost-effective way to assess the effectiveness of its asset management training efforts with USDA. The EPA official representing the drinking water needs assessment survey said that there would be value for the agency in asking water utilities about asset management.

Second, EPA and USDA do not collect information on the benefits and costs of using and implementing asset management to encourage use. EPA and USDA do not—and are not required to—compile information on the benefits and costs of implementing asset management to encourage small utilities to use it. EPA and USDA officials stated that increasing water and wastewater utilities' use of asset management increases the utilities' managerial and financial capacity and, for this reason, EPA and USDA (through technical assistance providers) share anecdotal data to encourage water utilities to adopt asset management. In particular, these officials said that they promote managerial and financial capacity building trainings, which include asset management training, by attending conferences at which they present to water utility officials, provide one-on-one technical assistance outreach, and engage in conversations with state SRF and USDA officials. EPA also provides some information about the potential benefits of asset management in documents available on its website. For example, EPA's 2008 best practices guide provides a list of the benefits of using asset management.

EPA and USDA's training materials for their workshops on *Rural and Small Systems Guidebook to Sustainable Utility Management* include examples of management challenges and best practices— many of which are asset management practices—to address the challenges. The training materials do not, however, include cases showing communities' use of asset management and the resulting benefits, including the costs that could result from implementing these best practices. According to an EPA official, most of the information on benefits, including cost savings, and costs comes from specific anecdotes and materials that technical assistance providers have developed and conveyed through their individual training and interactions with water utilities' staff or governing bodies. Some of EPA's technical assistance providers, such as the EFCs, have used information on the benefits of asset management to encourage water utility board members and city councils to adopt asset management.

However, the agencies have not compiled information about the benefits and costs of asset management into a single document that is more broadly available to water utilities. EPA and USDA officials told us that they had not considered compiling information about the benefits of asset management into one source, and they are not required to. However, providing information on benefits and costs to those who have not attended the agencies' trainings could help encourage them to adopt asset management practices. We noted in a September 2002 report that agencies use information dissemination as one of several tools to achieve goals for programs in which agencies do not act directly, but inform and persuade others to act to achieve a desired outcome.[59] Additionally, in a September 2005 report, we reported that agencies can evaluate their efforts using fact-based understandings of how their activities contribute to accomplishing the mission and broader results and to identify and increase the use of program approaches that are working well.[60] In this same report, we stated that agencies can adopt a number of practices to enhance the usefulness of information. One of these best practices for improving the use of information is to ensure that it is, among other things, relevant, accessible, and easy to use. For example, in 2006, EPA and the Federal Highway Administration collaboratively conducted a case study review of communities' experiences, including the benefits, of implementing asset management across multiple infrastructure sectors such as water, wastewater, and transportation.[61] According to the study, the purpose of the review was to provide, in one resource, relevant examples of how communities were responding to their infrastructure needs by using asset management practices.

An EPA official told us that, in 2006, the agency considered compiling information on cost savings by asking a question in a potential nationwide study of water utilities' use of asset management but did not pursue this study because, among other things, the agency did not have the resources to pay for it at that time. This official told us that EPA has instead engaged in activities that cost less than a nationwide study, such as developing case studies or other small-scale information collection efforts. A USDA official said that the agency would be open to exploring ways to collaborate on a study of the benefits and costs of asset management for small water utilities. As shown by the EPA and Federal Highway Administration study, EPA provided useful examples of the benefits for entities considering asset management by compiling and making broadly available information about the benefits and costs of asset management. Consistent with best practices for using performance information, compiling the information that EPA and USDA technical assistance providers share with water and wastewater utilities to document the benefits and costs of asset management could provide a resource for a broader audience of small water utilities that are considering using this approach. Both EPA and USDA officials said that they had developed several materials for small water utilities through their coordinated efforts and that a compilation of existing cases and examples of communities' use of asset management, its benefits, including cost savings, and costs could be useful in persuading some water utilities to use asset management.

CONCLUSION

EPA and USDA have provided millions of dollars of federal funding to help small water utilities increase their technical, managerial, and financial capacity to better meet the challenge of repairing and replacing the nation's aging water infrastructure and, also, provide safe and clean water to communities. Both agencies have identified asset management as a tool that water utilities can use to increase their ability to address current and future infrastructure needs.

EPA and USDA have played a significant role in encouraging and helping water utilities to implement asset management through funding conditions, training, and other resources. With 68,000 water utilities across the country, it is important to know which are using asset management and which are not. However, EPA and USDA do not collect information on water utilities' use of asset management, particularly from utilities that have taken part in agency training sessions on asset management. Existing data collection efforts such as EPA's needs assessment surveys may be a cost-effective means of doing this. By continuing to include questions in the clean water needs assessment survey and considering questions about water utilities' use of asset management to include in the drinking water needs assessment survey, EPA may have better assurance that it is collecting information in a cost-effective way to assess the results of its asset management training efforts with USDA. In addition, persuading elected officials and communities of the need for infrastructure investment is important, as is the need to use asset management to make investment decisions. EPA and USDA share the benefits and costs of asset management in various documents provided on their websites and through technical assistance providers. However, the agencies have not compiled the information on the benefits and costs of asset management, particularly the cost savings, into one source. A documented compilation of the benefits and costs of asset management, including the cost savings, consistent with performance management best practices, that is widely available to water utilities, may be helpful to EPA and USDA in encouraging a broader audience of small water utilities to consider adopting asset management.

RECOMMENDATIONS FOR EXECUTIVE ACTION

As EPA and USDA continue to consider ways to track and promote water utilities' implementation of asset management, we recommend the following:

- First, that the Administrator of EPA direct the Office of Groundwater and Drinking Water and Office of Wastewater Management to continue to include questions on water utilities' use of asset management in the clean water needs assessment and consider including questions about water utilities' use of asset management in future drinking water infrastructure needs assessment surveys.
- Second, that the Administrator of EPA, and the Secretary of USDA, through the Rural Development Agency, consider compiling into one document the existing cases and examples of the benefits and costs of asset management and widely share this information with water utilities.

AGENCY COMMENTS AND OUR EVALUATION

We provided the Administrator of EPA and the Secretary of USDA with a draft of this report for review and comment. In written comments provided by EPA, EPA generally agreed with our findings and recommendations. In addition, in an e-mail received from the GAO and OIG Liaison Officer within USDA's Rural Development agency, USDA agreed with our report.

In response to our recommendation that EPA continue to include questions on water utilities' use of asset management in the clean water needs assessment and consider including such questions in future drinking water infrastructure needs assessment surveys, EPA included "a significant caveat" to its agreement. Specifically, EPA's comments stated that the agency generally agrees with the recommendation, with the significant caveat that the method for continuing to assess the effectiveness of the agency's asset management training and technical assistance must be both effective and efficient. It further stated that although the agency has included asset management questions as part of the needs assessment surveys in the past, this mechanism has led to limited information regarding the level of implementation of asset management at utilities. Further, the comments stated that the needs assessment survey may not be the most efficient and effective way to collect these data since the survey's primary focus and design is to assess and quantify the nation's infrastructure need and not the adequate implementation of asset management. EPA stated that it would be willing to explore other means of obtaining data that would provide an indication of how utilities are benefitting from the agency's asset management training and technical assistance. We continue to consider the needs assessment survey to be a cost-effective and efficient method for collecting data from water utilities. If EPA explores others approaches and finds that the information can be systematically collected from the nation's water utilities for comparison over time, we agree other approaches could be appropriate.

In response to our recommendation that EPA and USDA consider compiling, and widely share, existing cases and examples of the benefits and costs of asset management, EPA noted that it agrees that it is important to educate utilities on the benefits of asset management in protecting the nation's infrastructure investment and described steps it has taken to do so. EPA stated that, as funding and resources allow, the agency would most likely consider the development of a case study compilation document focused on local decision makers who are key to ensuring that asset management is a priority and is implemented appropriately. We agree that focusing on local decision makers is important and believe that a document compiling case studies could be useful and made available to water utilities as well as local decision makers. USDA did not comment specifically on this recommendation and stated in its e-mail message that the agency will continue to emphasize asset management through its technical assistance providers funded through the agency's Technical Assistance and Grant program and joint EPA/USDA Work-Shop-in-a-Box initiative.

J. Alfredo Gómez
Director,
Natural Resources and Environment

APPENDIX I: REQUIREMENTS AND INCENTIVES FOR ASSET MANAGEMENT

Federal law does not require water utilities to implement asset management. However, in 2014, federal law began requiring all recipients of Clean Water State Revolving Fund (SRF) loans for repair, replacement, or expansion of a water utility to develop fiscal sustainability plans.[1] According to the law, fiscal sustainability plans should include (1) an inventory of critical assets; (2) an evaluation of the condition and performance of such assets or groups of assets; and (3) a plan for maintaining, repairing, and, as necessary, replacing the utility and a plan for funding such activities. According to Environmental Protection Agency (EPA) officials, some of the activities required as part of the fiscal sustainability plan are asset management practices.

Table 1. Reported Drinking Water and Clean Water SRF Asset Management Requirements and Incentives for the States We Reviewed

Drinking Water SRF				Clean Water SRF		
State	Required	Required for loan forgiveness	Priority points	Required	Required for loan forgiveness	Priority points
Arizona						
Arkansas			√			
Delaware						√
Idaho			√			√
Iowa						
Maine					√	
Minnesota						
Mississippi						
New Mexico			√	√		√
New York			√			
Wyoming						
Vermont			√			

Legend: √ = State officials we interviewed told us the state program has a requirement or incentive for asset management programs and/or plans. A blank cell indicates that state officials we interviewed told us that the state does not have a requirement or incentive for asset management. Sources: GAO analysis of interviews with state Drinking Water and Clean Water SRF officials. | GAO-16-237

Note: The states listed in this table are those states we included in our review. We did not ask water utilities about requirements for loan forgiveness in our interviews.

Officials in one state told us their state requires water utilities to develop asset management plans as a condition of SRF loans for water infrastructure, and officials in another state told us their state requires asset management plans for SRF loan forgiveness.[2] Other states may provide incentives during the application process. Specifically, states may award additional points in the application scoring process known as "priority points." Table 1 provides information about the requirements and incentives reported by officials in the states we reviewed.

APPENDIX II: OBJECTIVES, SCOPE, AND METHODOLOGY

This report examines selected water utilities' use of asset management. Our objectives were to examine (1) what is known about the use of asset management among the nation's water utilities—particularly small water utilities—including benefits and challenges, if any, for water utilities implementing asset management and (2) steps, if any, that the Environmental Protection Agency (EPA) and U.S. Department of Agriculture (USDA) are taking to help small water utilities implement asset management.

To examine what is known about water utilities' use of asset management, including benefits and challenges, if any, we used EPA's framework for asset management. The framework, from EPA's 2008 *Asset Management: A Best Practices Guide*,[1] describes five components and the practices that comprise asset management. The components and practices described in this document formed the basis of our interview questions. We also used other EPA documents describing the agency's asset management framework: 2003 *Asset Management: A Handbook for Small Water Systems: One of the Simple Tools for Effective Performance (STEP) Guide Series* [2] and 2008 *Building an Asset Management Team*[3]. This criterion remains relevant today because it is a federally-developed asset management framework for water utilities.

To determine if there were existing sources of information on the extent to which utilities use asset management, we interviewed EPA and USDA Rural Development staff about data each agency collects on water utilities' use of asset management, including any data on asset management collected by EPA in its national needs assessment surveys. We also interviewed representatives of national water and wastewater associations to identify potential data sources. Through this process, we identified one national study, a market research study conducted by McGraw-Hill Construction, a company that provides analytics, news, and intelligence for the North American construction industry, and CH2M, a company that, among other things, provides consulting services related to asset management.[4] The report described the results of a survey of 451 persons representing water utilities in the United States and Canada and the extent to which they had adopted 14 asset management practices. Thirty percent of the 451 persons in the survey sample represented utilities providing only drinking water services and 70 percent represented utilities providing drinking water and wastewater services. The 14 practices included such actions as (1) the use of a computerized maintenance management system, (2) the use of an asset register to facilitate analysis and planning, (3) the development of customer service and asset service-level performance measures, and (4) consideration of risks and consequences of alternative investment/ budget decisions, but the report did not describe how the 14 practices were selected. The report also included information from confidential interviews with water utilities on their insights in implementing asset management. We reviewed the authors' description of the study's methodology and determined that the data were sufficient for the purpose of describing qualitative information to corroborate information we obtained from our interviews about large utilities' use of asset management because the study authors included asset management practices that were similar to those identified by EPA.

To understand the extent to which small utilities are using asset management, we conducted semistructured interviews with officials in a nonprobability sample of 10 states: Arizona, Arkansas, Delaware, Idaho, Iowa, Minnesota, Mississippi, New York, Vermont, and Wyoming. To select these 10 states, we identified the state in each of EPA's 10 regions with the highest percentage of small water utility needs, using EPA's most recent needs assessment data from the 2011 *Drinking Water Infrastructure Needs Survey and Assessment* and the 2008 *Clean Watersheds Needs Survey*.[5] EPA's 2011 *Drinking Water Infrastructure Needs Survey and Assessment* calculated the need among small water utilities serving fewer than 10,000 people for 35 of the 50 states. For the remaining 15 states, EPA provided data on small water utilities serving fewer than 3,300 people. We calculated the percentages of small water utilities' share of statewide need for these 15 states using this information. The 2008 *Clean Watersheds Needs Survey* calculated the need for small water utilities serving fewer than 10,000 people for 47 states, the District of Columbia, and U.S. territories. It did not report data for 3 states, Alaska, North Dakota, or Rhode Island. As a result, we did not include these states in the data we used for our selection.

In our sample of 10 states, we used a standard set of questions for conducting interviews, by telephone, with state drinking water and clean water State Revolving Fund (SRF) program officials and USDA state office staff. Our standard set of questions consisted of closed- and open-ended questions to ensure we consistently captured officials' responses. During these interviews, we asked officials to estimate the use of asset management practices by water utilities' in their state, the benefits for utilities and lenders of using asset management, the challenges small utilities experience in implementing asset management, funding and technical assistance for asset management available to water utilities in their state, and the asset management practices for which small utilities are most in need of technical assistance. In addition, we asked USDA officials about the loan conditions they consider to be asset management practices. We also specifically asked SRF officials about requirements for asset management practices or plans as a condition of SRF loans. In these interviews with officials representing state SRF programs, USDA state offices, and small water utilities, we did not receive answers to every close-ended question we asked; we note in the report the number of answers provided for each question. Because our sample of states was a nonprobability sample, responses from the officials we interviewed cannot be generalized to other states and their water utilities, but they illustrate some of the uses of asset management practices among small water utilities in states with the greatest infrastructure needs.

In addition to interviewing state SRF and USDA officials in the 10 selected states, we interviewed, by telephone, officials in a nongeneralizable, random sample of small drinking water and wastewater utilities serving populations of 10,000 or less. To select these small water utilities, we used two EPA databases of water utilities. To identify drinking water utilities, we used EPA's Safe Drinking Water Information System, a database of information about drinking water utilities and their regulatory violations. To identify wastewater utilities, we used a publicly available database of the water utilities included in the 2008 *Clean Watersheds Needs Survey*. We assessed the data reliability for both databases by, among other things, reviewing published documents and data regarding EPA's quality assurance and quality control procedures for the needs survey assessment tools, contacting EPA officials to ensure we used the correct search fields and parameters, and reviewing past GAO reports and other documentation on the reliability of the data. Through these steps, we determined that the data were sufficiently reliable for our purposes of sampling drinking water and wastewater utilities in each of our 10 states. We selected a sample of 40 utilities, two drinking water and two wastewater utilities in each of the 10 selected states, and conducted interviews with at least one drinking water and one wastewater utility in 9 of the 10 states, for a total of 25 water utility interviews. (All of the water utilities we contacted in Mississippi declined to participate or did not respond to our interview requests.) To ensure our sample represented a range of small water utilities, we selected one drinking water utility serving a population of more than 500, but less than or equal to 10,000 and one drinking water utility serving a population of 500 or fewer people. Similarly, for each of the 10 states we selected one wastewater utility serving a population between 1,000 and 10,000 and one wastewater utility serving a population of less than 1,000 in each of the 10 selected states. In addition to the 40 utilities, we also generated a back-up list of randomly selected water utilities from which to choose if the utilities in the original sample declined to participate in our review or did not respond to our requests for an interview after three or more attempts. In total, we contacted 68 water utilities in all 10 of the states and conducted interviews with officials representing 25 water utilities in 9 states. Of these 25, 12 were drinking water utilities, and 13 were wastewater utilities. Table 2 provides a summary of the population served and ownership for the water utilities we interviewed in the 10 states.

Table 2. Water Utilities Interviewed in the 10 States, by Population Served and Ownership

	Population served	**Number of utilities**	**Ownership**	**Number of utilities**
Drinking water	Less than 500	5	Public	9
	501 to 10,000	7	Private	3
Wastewater	Less than 1,000	7	Public	13
	1,000 to 10,000	6	Private	0
Total		**25**		**25**

Source: GAO. | GAO-16-237

Our interviews with small water utilities consisted of a standard set of closed- and open-ended questions. Officials participating in these interviews were, for example, water utility operators or superintendents, maintenance staff, public works directors, elected city officials, water utility board members, and engineers. We asked about officials' familiarity with asset management as defined by EPA, the extent to which they were implementing asset management practices, and, if so, the costs and cost savings they had identified, whether water utility staff or governing officials had received technical assistance on asset management, and contextual background on the community in which they served. Because our sample of water utilities in the 10 selected states was a nongeneralizable sample, we do not use the data collected from these states to generalize about the use of asset management in other states and by other water utilities.

To analyze the open-ended questions in our surveys, we conducted several content analyses. Specifically, we conducted a content analysis to categorize the benefits (to water utilities and federal agencies) of asset management and the challenges small utilities face in implementing asset management practices. To identify categories in which to classify the open-ended responses, we examined the responses and used content analysis software to count the words officials used most frequently and identified broad groupings of concepts. We classified the responses to the open-ended questions on benefits to water utilities into the following categories: (1) planning, (2) financial, (3) awareness (of the system and assets), (4) management, (5) technical, (6) other benefits, and (7) unaware of benefits. We classified the responses to the open-ended questions on benefits to federal agencies into the following categories: (1) system and (2) lender. We classified the responses to the open-ended questions on challenges into the following categories: (1) financial, (2) human resources, (3) support, (4) education, and (5) other challenges.

Where appropriate, we also identified subcategories to classify responses.

To conduct the content analysis of responses, two analysts independently assigned officials' responses to one or more categories and compared their analyses. All initial disagreements regarding the categorizations of officials' responses were discussed and reconciled. The analysts then tallied the number of responses in each category. We tabulated the responses from closed-ended questions counts. To characterize officials' views we identified throughout this report, we defined modifiers to quantify officials' views. For example, "most" represents instances in which at least one state official in more than five states provided a response. The modifiers are as follows:

For state SRF and USDA state offices in the 10 selected states,

- "most" represents state SRF and USDA officials in more than 5 states,
- "half" represents state SRF and USDA officials in 5 states, and
- "some" represents state SRF and USDA officials in less than 5 states.

Table 3. Description of the Water Utilities Interviewed in Maine and New Mexico

	Name of utility	Utility type	Ownership type	Utility size	Population served
Maine					
	Farmington Water District	Drinking Water	Public	Small	4,050
	Portland Water District	Drinking Water	Public	Large	136,945
	Great Salt Bay Sanitary District	Wastewater	Public	Small	1,403
	Thomaston Pollution Control Department	Wastewater	Public	Small	2,562
	Ogunquit Sewer District	Wastewater	Public	Small	1,092
	City of Saco	Wastewater	Public	Small	7,810
New Mexico					
	Agua Sana User's Association	Drinking Water	Public	Small	1,200
	El Valle Water Alliance[a]	Drinking Water	Public	Small	1,732
	Rainsville Water and Sanitation District	Drinking Water and Wastewater	Public	Small	250
	Eldorado Area Water and Sanitation District	Drinking Water and Wastewater	Public	Small	7,350
	Albuquerque Bernalillo County Water Utility Authority	Drinking Water and Wastewater	Public	Large	601,983
	Entranosa Water and Wastewater Association	Drinking Water and Wastewater	Private	Small	8,500

Sources: GAO summary of in-person interviews, EPA's Safe Drinking Water Information System, and 2008 Clean Watersheds Needs Survey database and other water utility documents. | GAO-16-237

Note: Data for utility type and ownership type are from in-person interviews with officials at these utilities. Data on utility size and population served are from EPA's Safe Drinking Water Information System (drinking water utilities) and the database for the 2008 Clean Watersheds Needs Survey (wastewater utilities).

[a] El Valle Water Alliance is a consortium of 11 small water utilities, operating under one management structure.

We also conducted two in-person visits to Maine and New Mexico. We selected Maine and New Mexico based on recommendations from EPA and USDA officials in the agencies' headquarters, national water and wastewater associations, and technical assistance providers. In addition, we selected New Mexico because of the state's requirement that water utilities have an asset management plan as a condition of Clean Water SRF infrastructure funding. The officials that recommended Maine generally told us they did so because the state has long encouraged utilities to adopt asset management. During our visits to Maine and New Mexico, we interviewed a total of 12 small and large, public and private, water utilities. We selected these water utilities based on recommendations from state SRF officials and technical assistance providers in each state and, in the case of New Mexico, EPA regional office staff. Table 3 provides descriptive information about the water utilities we interviewed in Maine and New Mexico.

To examine the steps, if any, that EPA and USDA have taken to help small water utilities implement asset management, we reviewed EPA and USDA guidance, reports, training materials, and software tools available on asset management. One particular guidance we used was the 2011 Memorandum of Agreement on *Promoting Sustainable Rural Water and Wastewater Systems*, which describes EPA's and USDA's joint efforts to promote the technical, managerial, and financial capacity of small utilities and includes an emphasis on promoting asset management.[6] We then interviewed EPA and USDA officials to understand the actions they have taken, the funds they have spent, and the efforts they have made to coordinate on asset management activities. We also interviewed technical assistance providers funded by EPA and USDA to conduct trainings and one-on-one technical assistance on asset management. These technical assistance providers included the EPA-funded Environmental Finance Centers at the University of New Mexico, Wichita State University, Cleveland State University, and University of North Carolina at Chapel Hill and the National Rural Water Association and Rural Community Assistance Partnership's national offices and local affiliates in Maine and New Mexico. We compared the information we collected about the steps EPA and USDA have taken to key practices related to federal agencies' training efforts and collection and dissemination of information that we identified in previous reports.[7]

We conducted this performance audit from January 2015 to January 2016 in accordance with generally accepted government auditing standards. Those standards require that we plan and perform the audit to obtain sufficient, appropriate evidence to provide a reasonable basis for our findings and conclusions based on our audit objectives. We believe that the evidence obtained provides a reasonable basis for our findings and conclusions based on our audit objectives.

APPENDIX III: FEDERAL FUNDING FOR ASSET MANAGEMENT

The Environmental Protection Agency (EPA) and the U.S. Department of Agriculture (USDA) have funded asset management activities through their existing programs and, more recently, used some of this funding for collaborative efforts. Specifically, EPA officials told us that they first began funding asset management activities through the Drinking Water State Revolving Fund (SRF) program when the 1996 amendments to the Safe Drinking Water Act authorized states to use a certain percentage of their grants for such programs. According to USDA officials, USDA has contracted with national organizations that incorporated asset management training as part of their work in assisting small water utilities in managing and operating their facilities. With the 2011 issuance of a memorandum of agreement on sustainable infrastructure, EPA and USDA agreed to collaborate on training and to coordinate agency activities and financial assistance in areas that would increase technical, managerial, and financial capacity, including through the use of asset management.[1] As a result, EPA and USDA spending on asset management activities falls under various larger programmatic areas.

EPA's spending falls into the following three categories:

- **Small Systems Training and Technical Assistance Grants.** EPA funds asset management activities through this grant program to improve technical, managerial, and financial capacity of small drinking water and wastewater utilities. In fiscal year 2014, EPA provided an estimated $13.1 million in grants to support training and technical assistance for small utilities. Of the estimated $13.1 million, about $3 million was used to provide training and technical assistance to water utilities to improve financial and managerial capacity, which included asset management.[2] In December 2015, legislation was enacted to reauthorize the small systems training and technical assistance grants program for $15 million per year for fiscal years 2015 through 2020.[3]

- **Drinking Water SRF**. EPA also funds asset management training and technical assistance through the Drinking Water SRF. Subject to certain limitations, states may reserve a portion of these grants to fund various activities including training and technical assistance. States can spend 2 percent of their SRF grants to provide small water utilities with technical assistance; up to 4 percent for state program administration and technical assistance to water utilities of any size; up to 10 percent for the development of technical, managerial, and financial capacity, operator certification programs, and other activities; and up to another 15 percent for a variety of activities that can also include programs to develop technical, managerial, and financial capacity. Examples of assistance that states can provide with these funds include such activities as written guidance, one-on-one coaching, and online and classroom training that can include asset management.
- **Environmental Finance Center Grant Program.** EPA also funds asset management activities through its Environmental Finance Center Grant Program. EPA officials said that in fiscal year 2014 they provided a total of $1 million in grants to selected public, private universities or colleges, and nonprofit organizations to provide technical assistance to communities on a range of EPA priorities, including improving financial capacity.[4] This assistance included one-on-one technical assistance, workshops and other classroom trainings, and written guidance. EPA's Environmental Finance Center at the University of New Mexico has been working on asset management since 2003 and has provided training and technical assistance on the use of asset management since 2006.

USDA funds asset management activities through two programs:

- **Water & Waste Disposal Technical Assistance & Training Grant**s. This program provides grants to various nonprofit organizations for technical assistance on managerial topics, assistance with preparing loan applications, and helping water utilities to find solutions to problems in operating their facilities. USDA provides the grant funds for 1 year. In fiscal year 2014, USDA provided an estimated $19 million to nonprofit organizations. Nonprofit organizations can apply to provide services to one state or multiple states. USDA gives priority to certain types of applicants, including those that serve communities of fewer than 5,500 or fewer than 2,500, and those that will primarily provide "hands on" technical assistance and training to water utility managers and operators experiencing problems with operations and maintenance or management.

- **Circuit Rider Program—Technical Assistance for Rural Water Systems.**[5] Under this program, the National Rural Water Association—a training and technical assistance organization serving small communities—is contracted to provide staff in each of the 50 states with technical assistance on day-to-day operational, managerial, and financial issues. Specifically, according to the information the National Rural Water Association publishes on its website,[6] staff known as "circuit riders" work on site with water utility personnel to troubleshoot problems, evaluate alternative technological solutions, recommend operational improvements, assist with leak detection, respond to natural disasters and other emergencies, provide hands-on training, participate in board and council meetings, and conduct user rate analyses. In fiscal year 2014, USDA provided about $15 million for the Circuit Rider Program.

End Notes

[1] American Society of Civil Engineers, *2013 Report Card for America's Infrastructure* (Reston, VA: American Society of Civil Engineers, 2013).

[2] EPA's most recent water infrastructure needs assessments estimated that funding needs for drinking water infrastructure totaled $384.2 billion (as of 2011) and for wastewater infrastructure totaled $271 billion (as of 2012). EPA conducts a separate survey and assessment for each type of infrastructure, drinking water and wastewater, on separate 4-year schedules. These cost estimates reflect the 20-year projected drinking water and wastewater infrastructure costs starting with the year that each survey was conducted. See *EPA, Drinking Water Infrastructure Needs Survey and Assessment: Fifth Report to Congress*, EPA 816-R-13-006 (Washington, D.C.: April 2013) and *EPA, Clean Watersheds Needs Survey 2012: Report to Congress*, EPA-832-R-15005 (Washington, D.C.: Jan. 2016). Throughout this report, all dollar figures, such as these costs, are expressed in nominal dollars, unadjusted for inflation.

[3] EPA uses the term "small" to refer to water utilities serving populations of 10,000 or less, and the U.S. Department of Agriculture uses "rural." Throughout this report, we use the term "small" to refer to these water utilities.

[4] We use the term "water utilities" to refer to both drinking water and wastewater utilities.

[5] The Safe Drinking Water Act applies to certain public water systems. Safe Drinking Water Act, Pub. L. No. 93-523, 88 Stat. 1660 (1974)(codified as amended at 42 U.S.C. §§ 300f–300j-26 (2015)). The Clean Water Act applies to all wastewater treatment. The Federal Water

Pollution Control Act Amendments of 1972, Pub. L. No. 92-500, 86 Stat. 816, codified as amended at 33 U.S.C. §§ 1251-1387 (2015) (commonly referred to as the Clean Water Act).

[6] EPA's Drinking Water SRF program was created under the Safe Drinking Water Act Amendments of 1996. Pub. L. No. 104-182, § 130, 110 Stat. 1613, 1662–1672 (codified as amended at 42 U.S.C. § 300j-12 (2015)).

[7] EPA's Clean Water SRF program was created under the Water Quality Act of 1987. Pub. L. No. 100-4, § 212, 101 Stat. 7, 21-28 (codified as amended at 33 U.S.C. §§ 1381–1388 (2015)). The Water Quality Act of 1987 amended the Clean Water Act.

[8] USDA's Water and Waste Disposal program was created under the Consolidated Farmers Home Administration Act of 1961, Amendments. Pub. L. No. 89-240, § 1, 79 Stat. 931, 931-932 (codified as amended at 7 U.S.C. § 1926 (2015)).

[9] Congressional Research Service, *Water Infrastructure Needs and Investment: Review and Analysis of Key Issues.* RL31116, (Washington, D.C.: Dec. 28, 2012).

[10] GAO, *Water Infrastructure: Comprehensive Asset Management Has Potential to Help Utilities Better Identify Needs and Plan Future Investments*, GAO-04-461 (Washington, D.C.: Mar. 19, 2004).

[11] The International Organization for Standardization (ISO) published a general international standard on the use of asset management.

[12] 23 U.S.C. 119(e)(1) (2015).

[13] GAO, *Water Infrastructure: Information on Financing, Capital Planning, and Privatization*, GAO-02-764 (Washington, D.C.: Aug. 16, 2002).

[14] EPA and USDA, Memorandum of Agreement Between the United States Environmental Protection Agency and the United States Department of Agriculture-Rural Development Rural Utilities Service, *Promoting Sustainable Rural Water and Wastewater Systems*, (Washington, D.C.: June 2011).

[15] An EPA request for grant applications provides definitions for technical, managerial, and financial capacity. Technical capacity refers to the physical infrastructure of the water utility and the ability of personnel to adequately operate and maintain the system and to apply the necessary technical knowledge (for example, knowledge necessary for certification) to comply with the law and regulations. Managerial capacity refers to the management structure and practices of the utility, for example, ownership accountability, staffing, and communication with customers and regulators. Financial capacity refers to the financial resources of the utility, such as having sufficient revenue, setting user rates, and operation budget and planning.

[16] At the time of our state selection, EPA had not released its 2012 report. Therefore, we used data associated with the EPA's 2008 Clean Watersheds Needs Survey. EPA, *Clean Watersheds Needs Survey 2008: Report to Congress*, EPA-832-R-10-002 (Washington, D.C.: May 2010).

[17] McGraw-Hill Construction, *Water Infrastructure Asset Management: Adopting Best Practices to Enable Better Investments: Smart Market Report, (Bedford, MA: 2013).* The McGraw-Hill Construction study was conducted in partnership with CH2M, a company that, among other things, provides asset management consulting services. McGraw-Hill Construction provides data, analytics, news, and intelligence for North America. The report described the results of a survey of 451 water utilities in the United States and Canada and the extent to which they had adopted 14 asset management practices. The study also included information from confidential interviews with water utilities on their experiences in implementing asset management.

[18] American Water Works Association, *2015 Establishing the Level of Progress in Utility Asset Management Survey Results*," (Denver, CO: Dec. 2015).

[19] Environmental Protection Agency, *Asset Management: A Best Practices Guide*, EPA 816-F-08-014 (Washington, D.C.: April 2008). Hereafter, we refer to this document as "EPA's 2008 best practices guide."

[20] Not all of the small water utilities we contacted responded to our request for an interview.

[21] The technical assistance providers we met with were nongovernmental organizations that provided a range of services to help small communities with regulatory, managerial, and financial needs related to their water utilities, such as training, assistance with completing loan applications, or studies of water rates.

[22] GAO, *Human Capital: Selected Agencies' Experiences and Lessons Learned in Designing Training and Development Programs*, GAO-04-291 (Washington, D.C.: Jan. 30, 2004); *Human Capital: A Guide for Strategic Training and Development Efforts in the Federal Government*, GAO-04-546G (Washington, D.C.: March 2004); *Managing for Results: Enhancing Agency Use of Performance Information for Management Decision Making*, GAO-05-927 (Washington, D.C.: Sept. 9, 2005); *Program Evaluation: Strategies for Assessing How Information Dissemination Contributes to Agency Goals*, GAO-02-923 (Washington, D.C.: Sept. 30, 2002).

[23] There are other frameworks and standards for asset management. In 2006, the Water Environment Research Foundation, with funding from EPA and the Water Research Foundation, developed the Sustainable Infrastructure Management Program Learning Environment in 2006 for use by water utilities. Standards include the Publicly Available Specification (PAS) 55, developed in 2004 by the UK-based Institute of Asset Management and British Standards Institution, and the ISO 55000 series, developed by the International Organization for Standardization in 2014. The International Organization for Standardization developed the ISO 55000 series from the PAS 55 and introduced the ISO 55000 series as an international standard in January 2015. Both the PAS 55 and ISO 55000 series are designed for multiple sectors.

[24] EPA, *Asset Management: A Best Practices Guide*, EPA 816-F-08-014 (Washington, D.C.: April 2008).

[25] In 2008, EPA issued a series of three fact sheets on the topics of best practices for implementing asset management, building an asset management team, and information about asset management for local officials.

[26] Environmental Protection Agency, *Building an Asset Management Team*, EPA 816-F08-016 (Washington, D.C.: April 2008).

[27] GAO-04-461.

[28] U.S. Conference of Mayors, *National City Water Survey 2007: The Status of Asset Management Programs in Public Water and Sewer Infrastructure in America's Major Cities* (Washington, D.C.: September 2007).

[29] Environmental Protection Agency, *Asset Management for Local Officials*, EPA 816-F-08- 015 (Washington, D.C.: April 2008).

[30] The name of USDA's Circuit Rider program reflects the delivery of the technical assistance—that is, staff with expertise in operating water utilities travel to provide targeted assistance to water utilities.

[31] Environmental Protection Agency, *National Characteristics of Drinking Water Systems Serving 10,000 or Fewer People*, EPA 816-R-10-022 (Washington, D.C.: July 2011).

[32] According to a 2002 Congressional Budget Office (CBO) report on drinking water and wastewater infrastructure financing, 50 percent of all drinking water utilities that serve the

public are privately-owned, and about 20 percent of wastewater utilities that serve the public are privately-owned.

[33] The methods that EPA uses to collect information about drinking water and wastewater utilities' needs differs by needs assessment and water utility size. To identify 20-year needs for drinking water utilities for the 2011 needs assessment survey, for example, EPA administered a questionnaire to all large water utilities and a sample of medium-sized water utilities in each state. EPA used data from the 2007 survey to estimate needs for small water utilities. To identify 20-year needs for wastewater utilities for the 2008 and 2012 needs assessment surveys, for example, states collect needs data from wastewater utilities and submit detailed information and supporting documents to EPA. The 2008 and 2012 *Clean Watersheds Needs Surveys* did not include information about the infrastructure needs of privately-owned wastewater utilities.

[34] EPA's needs assessment surveys are subject to federal law regarding the collection of information. Under the Paperwork Reduction Act of 1980, a federal agency must, among other things, submit for the approval of the OMB any proposal to collect the same information from 10 or more members of the public. The federal agency seeking approval must describe the information the agency seeks to collect from the public, the reason the information is needed, and estimates of the burden—the time, effort, and cost—imposed on the public to provide the information.

[35] According to EPA's website, a capital improvement plan assesses which projects (including asset improvements, repairs, replacements and such) need to be completed in the future.

[36] EPA, *Drinking Water Infrastructure Needs Survey and Assessment: Fifth Report to Congress.*

[37] EPA, *Clean Watersheds Needs Survey 2008: Report to Congress*. EPA's 2012 *Clean Watersheds Needs Survey* did not make statements about asset management.

[38] GIS are systems of computer software, hardware, and data used to capture, store, manipulate, analyze, and graphically present a potentially wide array of geospatial data. The primary function of a GIS is to link multiple sets of geospatial data and display the combined information as maps with different layers of information.

[39] Environmental Protection Agency, EPA 816-R-10-022.

[40] USDA defines short-lived assets as those with an estimated useful life of 15 years or less.

[41] Officials in one state told us that the large water utilities in their state were as likely as small water utilities to implement asset management.

[42] McGraw-Hill Construction's study defined a "small utility" as one serving less than 50,000 people and did not include water utilities serving populations of fewer than 3,300.

[43] Generally, "repairs" are defined as work going beyond normal maintenance and "replacement" is the substitution with a new or equivalent item.

[44] Subject to certain limitations states may use a portion of their drinking water SRF grants to fund various activities, including training, technical assistance and programs to develop technical, managerial, and financial capacity.

[45] The SRF programs are implemented by the states. While EPA establishes high-level requirements for the programs, states may impose additional or more specific requirements that differ by state.

[46] Under the Safe Drinking Water Act (SDWA), states may not provide Drinking Water SRF loan assistance to systems that do not have the technical, managerial, and financial capability to ensure compliance with the SDWA unless the use of such assistance will ensure compliance, and the system has agreed to make the necessary changes in operation to ensure that it has the technical, managerial, and financial capacity to comply over the long term.

[47] A preliminary engineering report describes the proposed project, including its purpose, features of the proposed location, condition of any existing facilities, alternative approaches considered, design features, and costs. The report is prepared using a template.

[48] The Water and Waste Disposal program is implemented by USDA under federal regulations that apply to all states.

[49] Officials in 5 of the 25 small water utilities we interviewed said that they had both an SRF and USDA loan.

[50] 7 C.F.R. §§ 1780.33, 1780.55 (2015).

[51] According to an EPA official, the intent of the SRF program is to provide states with flexibility in implementing their SRF programs.

[52] EPA provides grants through its EFC Grant Program to selected public and private universities or colleges and nonprofit organizations to provide technical assistance to communities on a range of EPA priorities, including improving financial capacity. This assistance includes studies of user rates for water utilities, one-on-one technical assistance, workshops and other classroom trainings, and written guidance. For fiscal years 2016 through 2021, EPA will provide grants to nine EFCs that serve 9 of its 10 regions.

[53] The National Rural Water Association is a nonprofit organization dedicated to training, supporting, and promoting the water and wastewater professionals that serve small communities across the United States.

[54] The Rural Community Assistance Partnership is a federally-funded network of nonprofit organizations that provide training, technical assistance, and other resources related to water and wastewater issues for small, rural communities and tribes. Most of its project communities are economically disadvantaged and have a population under 2,500.

[55] EPA and USDA, *Rural and Small Systems Guidebook to Sustainable Utility Management,* (Washington, D.C.: 2013). According to EPA and USDA officials and EPA's website, the guidebook is adapted from another guidebook and training course, "Effective Utility Management," designed for medium to large utilities.

[56] USDA and EPA, *Making A Difference for Rural and Small Utilities: Sustainable Rural and Small Utility Management Initiative*, (Washington, D.C.: 2015).

[57] GAO-04-291 and GAO-04-546G.

[58] GAO-05-927.

[59] GAO-02-923.

[60] GAO-05-927, 15.

[61] Environmental Protection Agency and Department of Transportation, Federal Highway Administration, *Multisector Asset Management Case Studies* (Washington, D.C.).

End Notes for Appendix I

[1] Water Resources Reform and Development Act of 2014, Pub. L. No. 113-121, § 5003, 128 Stat. 1193, 1323 (codified at 33 U.S.C. § 1383(d)(1)(E)(i) (2015).

[2] We interviewed state officials responsible for managing EPA's Drinking Water and Clean Water SRF programs as well as officials in USDA's state offices in a nonprobability sample of 10 states. To select these states, we used EPA's most recent needs assessment data (i.e., 2011 for drinking water infrastructure and 2008 for wastewater infrastructure) to identify the state in each of EPA's 10 regions with the highest percentage of small water utility needs. These states were: Arizona, Arkansas, Delaware, Idaho, Iowa, Minnesota,

Mississippi, New York, Vermont, and Wyoming. See *EPA, Drinking Water Infrastructure Needs Survey and Assessment: Fifth Report to Congress*, EPA 816-R-13- 006 (Washington, D.C.: April 2013) and *Clean Watersheds Needs Survey 2008: Report to Congress*, EPA-832-R-10-002 (Washington, D.C.: May 2010). At the time of our state selection, EPA had not released its 2012 report.

End Notes for Appendix II

[1] Environmental Protection Agency, Asset Management: A Best Practices Guide, EPA 816-F-08-014 (Washington, D.C.: April 2008).

[2] Environmental Protection Agency, Building an Asset Management Team, EPA 816-F-08- 016 (Washington, D.C.: April 2008).

[3] Environmental Protection Agency, Asset Management: A Handbook for Small Water Systems: One of the Simple Tools for Effective Performance (STEP) Guide Series, EPA 816-R-03-016, (Washington, D.C.: September 2003).

[4] McGraw-Hill Construction, Water Infrastructure Asset Management: Adopting Best Practices to Enable Better Investments: Smart Market Report (Bedford, MA: 2013).

[5] Environmental Protection Agency, Drinking Water Infrastructure Needs Survey and Assessment: Fifth Report to Congress, EPA 816-R-13-006 (Washington, D.C.: April 2013) and Environmental Protection Agency, Clean Watersheds Needs Survey 2008: Report to Congress, EPA-832-R-10-002 (Washington, D.C.: May 2010).

[6] EPA and USDA, Memorandum of Agreement Between the United States Environmental Protection Agency and the United States Department of Agriculture-Rural Development Rural Utilities Service, Promoting Sustainable Rural Water and Wastewater Systems (Washington, D.C.: June, 2011).

[7] GAO-04-291, GAO-04-546G, GAO-05-927, and GAO-02-923.

End Notes for Appendix III

[1] EPA and USDA, Memorandum of Agreement Between the United States Environmental Protection Agency and the United States Department of Agriculture-Rural Development Rural Utilities Service, *Promoting Sustainable Rural Water and Wastewater Systems* (Washington, D.C.: June, 2011).

[2] According to an EPA document summarizing the grant program and fiscal year 2014 awards, the remaining estimated $10.1 million was used to assist operators and owners of small public water systems in understanding requirements of the Safe Drinking Water Act (SDWA) and specific National Primary Drinking Water Regulations; and, training to help owners of private wells improve water quality.

[3] Grassroots Rural and Small Community Water Systems Assistance Act, Pub. L. No. 114- 346, § 4 (2015).

[4] According to EPA documents, the agency requested $2 million in grant funds for this program for fiscal year 2016.

[5] The name of USDA's Circuit Rider program reflects the delivery of the technical assistance—that is, staff with expertise in operating water utilities travel to provide targeted assistance to water utilities.

[6] National Rural Water Association, http://nrwa.org/initiatives/training-and-technicalassistance/ (accessed on November 24, 2015).

In: Water Utilities Use ...
Editor: Alfred Caldwell
ISBN: 978-1-63485-937-0

Chapter 2

WATER INFRASTRUCTURE: COMPREHENSIVE ASSET MANAGEMENT HAS POTENTIAL TO HELP UTILITIES BETTER IDENTIFY NEEDS AND PLAN FUTURE INVESTMENTS*

United States General Accounting Office

WHY GAO DID THIS STUDY

Having invested billions of dollars in drinking water and wastewater infrastructure, the federal government has a major interest in protecting its investment and in ensuring that future assistance goes to utilities that are built and managed to meet key regulatory requirements. The Congress has been considering, among other things, requiring utilities to develop comprehensive asset management plans. Some utilities are already implementing asset management voluntarily. The asset management approach minimizes the total cost of buying, operating, maintaining, replacing, and disposing of capital assets during their life cycles, while achieving service goals. This report discusses (1) the benefits and challenges for water utilities in implementing

* This is an edited, reformatted and augmented version of a United States General Accounting Office, Publication No. GAO-04-461, dated March 2004.

comprehensive asset management and (2) the federal government's potential role in encouraging utilities to use it.

WHAT GAO RECOMMENDS

Among other things, GAO is recommending that the Environmental Protection Agency (EPA) (1) better coordinate its own activities to facilitate information sharing and reduce the potential for duplication and (2) ensure that water utilities have access to information they can use by establishing a Web site focused on asset management. In commenting on a draft of this report, EPA generally agreed with the report and its recommendations.

WHAT GAO FOUND

Drinking water and wastewater utilities that GAO reviewed reported benefiting from comprehensive asset management but also finding certain challenges. The benefits include (1) improved decision making about their capital assets and (2) more productive relationships with governing authorities, rate payers, and others. For example, utilities reported that collecting accurate data about their assets provides a better understanding of their maintenance, rehabilitation, and replacement needs and thus helps utility managers make better investment decisions. Among the challenges to implementing asset management, utilities cited collecting and managing needed data and making the cultural changes necessary to integrate information and decision making across departments. Utilities also reported that the shorter-term focus of their governing bodies can hamper long-term planning efforts.

EPA currently sponsors initiatives to promote the use of asset management, including educational materials, technical assistance, and research. While this is a good first step, GAO found that EPA could better coordinate some activities. For example, EPA has no central repository to facilitate information sharing within and across its drinking water and wastewater programs, which would help avoid duplication of effort. Water industry officials see a role for EPA in promoting asset management as a tool to help utilities meet infrastructure-related regulatory requirements; they also noted that establishing an EPA Web site would be useful for disseminating asset management information to utilities. The officials raised concerns, however, about the implications of mandating asset management, citing challenges in defining an adequate asset management plan and in the ability of states to oversee and enforce compliance.

ABBREVIATIONS

EPA	Environmental Protection Agency
GAO	General Accounting Office

* * *

March 19, 2004
The Honorable James M. Jeffords
Ranking Minority Member
Committee on Environment and Public Works
United States Senate

Dear Senator Jeffords:

In response to your request, this report examines (1) the potential benefits of comprehensive asset management for drinking water and wastewater utilities and the challenges that could hinder its implementation and (2) the role that the federal government might play in encouraging utilities to implement asset management.

Sincerely yours,

John B. Stephenson
Director, Natural Resources and Environment

EXECUTIVE SUMMARY

Purpose

Mounting evidence suggests that the integrity of the nation's drinking water and wastewater infrastructure is at risk without a concerted effort to improve the management of key assets—pipelines, treatment plants, and other facilities—and a significant investment in maintaining, rehabilitating, and replacing these assets. According to recent studies by the Environmental Protection Agency (EPA) and other organizations, drinking water and wastewater utilities will need to invest hundreds of billions of dollars in their capital infrastructure over the next two decades. However, if utilities maintain current spending levels, financing the needed investments could be problematic. Based on a survey of several thousand drinking water and wastewater utilities, GAO reported in August 2002 that a significant percentage of the utilities—29 percent of the drinking water utilities and 41 percent of the wastewater utilities—were not generating enough revenue from user rates and other local sources to cover their full cost of service.[1] Furthermore, roughly one-third of the utilities (1) deferred maintenance because of insufficient funding, (2) had 20 percent or more of their pipelines nearing the end of their useful life, and (3) lacked basic plans for managing their capital assets.

Each year, the federal government makes available billions of dollars to help local communities finance drinking water and wastewater infrastructure projects. However, concerns about the condition of existing infrastructure have prompted calls to increase this financial assistance and, at the same time, ensure that the federal government's investment is protected. The Congress has been considering a number of infrastructure-related proposals, including requirements for local utilities to have asset management plans. Some utilities have turned to comprehensive asset management on their own initiative. This approach to managing capital infrastructure focuses on minimizing the total cost of acquiring, operating, maintaining, replacing, and disposing of capital assets over their life cycle and doing so in a way that achieves the level of service customers desire. Among other things, comprehensive asset management allows utility managers to obtain better information on the age and condition of existing assets, determine the level of maintenance needed to optimize asset performance and useful life, assess the risks associated with the failure of various assets and set priorities for their maintenance and replacement, understand the trade-offs and implications of management decisions about the assets, and use better information to justify proposed rate increases or capital investments. Water industry officials agree that by making informed decisions about investments in capital assets, drinking water and wastewater utilities can better justify the rate increases associated with making needed improvements to their infrastructure.

The Ranking Minority Member of the Senate Committee on Environment and Public Works asked GAO to examine (1) the potential benefits of comprehensive asset management for drinking water and wastewater utilities and the challenges that could hinder its implementation and (2) the role that the federal government might play in encouraging utilities to implement asset management.

To address the first issue, GAO conducted initial interviews with 46 U.S. drinking water and wastewater utilities identified by knowledgeable government and water industry officials as implementing comprehensive asset management. To obtain more detailed information on the benefits and challenges associated with implementing asset management, GAO conducted structured interviews with 15 of these utilities, selected because they (1) reported or anticipated achieving quantitative benefits from asset management or (2) represented smaller entities. Our results are not generalizable to the larger universe of domestic drinking water and wastewater utilities. Because asset management is a relatively new concept for domestic water utilities, GAO supplemented the structured interviews by obtaining information from 6 utilities and five government agencies in Australia and New Zealand—two countries that have taken the lead in implementing comprehensive asset management in their drinking water and wastewater utilities—and two private companies with longstanding experience in using comprehensive asset management in their respective fields, which provided additional information on benefits and challenges.

To address the second issue, GAO discussed options for a federal role in promoting asset management with the 15 utilities selected for structured interviews, water industry associations, and EPA. In addition, based on contacts with a variety of organizations and officials experienced in asset management, GAO identified the U.S. Department of Transportation as being at the forefront of federal involvement in this issue. GAO obtained information about the department's initiatives from the Office of Asset Management, within the Federal Highway Administration.

We conducted our work between March 2003 and March 2004 in accordance with generally accepted government auditing standards.

BACKGROUND

At its most basic level, comprehensive asset management involves the systematic collection of key data and the application of analytical tools such as life-cycle cost analysis and risk assessment. Asset management thus provides information that managers can use to make sound decisions about their capital assets and allows decision makers to better identify and manage needed investments in their organization's infrastructure. By following this approach, organizations also change the process they use to make decisions, including the types of information they bring to bear and which segments of the organization participate in the decision-making process. Using a fully integrated decision process, many segments of an organization, including accounting, engineering, finance, maintenance, and operations, are expected to exchange relevant information, share in the decision making, and take an organizationwide view when setting goals and priorities. For drinking water and wastewater utilities, an integral part of a comprehensive asset management program is ensuring that adequate funds are available through user rates or other means so that asset management decisions can be implemented (e.g., ensuring that planned maintenance can be conducted and capital assets can be repaired, replaced, or upgraded on schedule).

Comprehensive asset management is a relatively new concept for drinking water and wastewater utilities in the United States. Domestic utilities implementing asset management are generally large and vary considerably in terms of their approach. For example, some are applying the concepts of asset management on a utilitywide basis and others are beginning in specific departments or facilities. In implementing asset management, domestic utility managers have drawn from the experiences of international utilities that are considered to be at the forefront of asset management for drinking water and wastewater infrastructure. For example, in Australia and New Zealand, where the concept has been strongly endorsed by the national governments, water utilities have used comprehensive asset management for about 10 years. In each country, a key impetus for the move toward asset management was legislation that called for water utilities to improve their financial management, requiring utilities to recover the full cost of service in Australia and, in New Zealand, to annually depreciate their assets and use cost-benefit analysis.

In the United States, the Congress has been considering proposals to require utilities to adopt key components of asset management, such as inventorying critical assets, evaluating their condition and performance, and developing plans to (1) maintain, repair, and replace assets and (2) fund such activities. These proposals typically link the use of asset management to a utility's eligibility for federal financial assistance in making infrastructure improvements.

The universe of drinking water and wastewater utilities indicates that efforts to adopt asset management may vary considerably because of the utilities' sizes and their ability to marshal resources for the effort. In the United States, about 54,000 community water systems supply most of the nation's drinking water and about 16,000 wastewater treatment systems provide sewer service. Larger utilities account for much of the projected infrastructure needs; for example, drinking water systems serving more than 10,000 people account for approximately 65 percent of the estimated needs for such utilities. However, most utilities are small, with 93 percent of community drinking water systems and 71 percent of wastewater systems serving 10,000 people or fewer. EPA has found that smaller utilities are less likely to have the financial, managerial, and technical capacity to meet regulatory requirements and are less likely to cover their full cost of providing services.

RESULTS IN BRIEF

U.S. drinking water and wastewater utilities that GAO contacted reported benefiting from applying the concepts of asset management to their operations but have also encountered certain challenges. Utilities are seeing benefits from (1) improved decision making because they have more accurate and integrated information about their capital assets and (2) more productive relationships with governing authorities, ratepayers, and other stakeholders because they can provide better information in a more transparent way. For example, utilities reported that collecting accurate data about their assets provides a better understanding of their maintenance, rehabilitation, and replacement needs, which helps utility managers make better investment decisions. While water industry officials identified financial and other benefits from using asset management, reported savings should be interpreted carefully. According to the utilities that GAO contacted, the challenges associated with implementing asset management included collecting and managing needed data and making the cultural changes necessary to integrate information and decision making across departments. Utilities also reported that the shorter-term focus of their governing bodies can hamper long-term planning efforts.

Although smaller utilities face more obstacles to implementing asset management, largely as a result of limited resources, such utilities can also benefit from applying asset management concepts.

EPA can play a stronger role in encouraging water utilities to use asset management by leveraging ongoing efforts within and outside the agency.

EPA currently sponsors initiatives to promote the use of asset management, such as developing educational materials; providing technical assistance, particularly to smaller utilities; and funding research related to asset management. Nevertheless, GAO found that some activities could be better coordinated. For example, EPA's Office of Ground Water and Drinking Water and Office of Wastewater Management fund state and university-based training and technical assistance centers that have developed guidance manuals, tools, and training materials to assist small utilities with asset management. However, no central repository exists to facilitate information sharing within and across the drinking water and wastewater programs and thereby avoid duplication of effort. GAO also found that opportunities exist for EPA to coordinate with other federal agencies, such as the Department of Transportation, that have already developed tools and training materials on asset management. When asked for their views on a potential federal role in asset management, water industry officials said that EPA should assume a greater leadership role in promoting asset management. For example, they see asset management as a tool to help utilities meet regulatory requirements that depend on maintaining an adequate infrastructure and believe that EPA should establish a Web site to serve as a central repository of information on implementing asset management. However, the officials raised concerns about the implications of mandating asset management, citing challenges in defining an adequate asset management plan and in the ability of states to oversee and enforce compliance. GAO is making recommendations to strengthen EPA's existing efforts to promote water utilities' use of asset management.

Principal Findings

Utilities See Benefits from Using Comprehensive Asset Management, but Face Implementation Challenges

Utilities that have started using comprehensive asset management report benefits for several aspects of their operations. For example, collecting, sharing, and analyzing data on capital assets has allowed utilities to make more informed decisions about how best to manage the assets. In particular, utilities are using the information they collect to allocate their maintenance resources more effectively and make better decisions about whether to rehabilitate or replace aging assets. Utilities also report that asset management fosters information sharing across departments as well as coordinated planning and decision making. These improvements help utility managers reduce duplication of effort and improve the allocation of staff time and other resources.

Utilities also report that comprehensive asset management benefits their relations with external stakeholders, such as local governing bodies, ratepayers, and credit rating agencies. For instance, several utilities have used, or expect to use, the information collected through comprehensive asset management to persuade elected officials to increase user rates to help pay for needed improvements in drinking water and wastewater infrastructure. Although water industry officials identified financial and other benefits from using asset management, reported savings should be viewed with caution because, for instance, comprehensive asset management may be implemented concurrently with other changes in management practices or operational savings may be offset by increases in capital expenditures.

Despite the acknowledged benefits of comprehensive asset management, utilities report several key challenges that can hinder efforts to implement this approach. For example, collecting the appropriate data on utility assets and managing the information efficiently can be difficult when existing data are incomplete and inaccurate or the data come from multiple departments and are maintained using different and incompatible software programs. Utilities reported that another major challenge is overcoming resistance to cultural change and fostering more communication among departments that do not regularly exchange information. Utility officials believe that it is essential to change the management culture to encourage more interdepartmental coordination and information sharing. Finally, although asset management provides utilities with better information to justify needed rate increases, their justifications may not be effective because of pressure to keep rates low and competing priorities for local revenues. Utility and water industry officials cited the difficulty of trying to implement long-term capital improvement plans when governing bodies have a shorter-term focus as a key challenge to asset management. Smaller utilities may have more difficulty implementing asset management because they typically have fewer financial, technological, and staff resources. On the other hand, because such utilities have fewer capital assets to manage, they can turn to low-cost management alternatives that do not require expensive or sophisticated technology.

EPA Can Promote the Use of Asset Management By Strengthening Existing Initiatives

EPA currently sponsors initiatives to encourage the use of comprehensive asset management through its partnerships with water industry associations and state and university-based training and technical assistance centers. These initiatives include developing training and informational materials, providing technical assistance, and funding research related to asset management. While this is a good first step, GAO found that better coordination of these efforts within and across the drinking water and wastewater programs could reduce the potential for duplication of effort and help ensure that limited resources are used effectively. GAO also found opportunities for EPA to leverage its resources by adapting the asset management tools and informational materials available from other federal agencies with experience in asset management, such as the Federal Highway Administration in the Department of Transportation.

When asked for their views on a potential federal role, water industry and utility officials said that given the benefits of asset management, it is in EPA's interest to assume a greater leadership role in promoting its use. For example, the officials indicated that one useful option would be educating utilities about the potential for asset management to help them comply with certain regulatory requirements that focus to some degree on the adequacy of utility infrastructure and the management practices that affect it. As a case in point, the officials cited requirements for ensuring that drinking water utilities have the financial, managerial, and technical capacity they need to provide safe drinking water over the long term. GAO also found support for an EPA Web site that would serve as a central repository of information on comprehensive asset management and provide drinking water and wastewater utilities with direct and easy access to implementation tools and training materials developed by EPA and others. Water industry associations and individual utilities questioned the feasibility of proposed requirements for asset management plans, particularly as a condition of receiving federal financial assistance. While the proposals are consistent with what GAO has found to be the leading practices in capital decision making,[2] the officials expressed concerns about, among other things, (1) whether state regulators have the resources to assess the adequacy of asset management plans and oversee compliance and (2) the potential for a mandate to limit the flexibility utilities need to tailor asset management to their individual circumstances.

RECOMMENDATIONS FOR EXECUTIVE ACTION

Given the potential of comprehensive asset management to help water utilities better identify and manage their infrastructure needs, the Administrator, EPA, should take steps to strengthen the agency's existing initiatives on asset management and ensure that relevant information is accessible to those who need it. Specifically, the Administrator should

- better coordinate ongoing and planned initiatives to promote asset management within and across the drinking water and wastewater programs to leverage limited resources and reduce the potential for duplication;
- explore opportunities to take advantage of asset management tools and informational materials developed by other federal agencies;
- strengthen efforts to educate utilities on how implementing asset management can help them comply with certain regulatory requirements that focus in whole or in part on the adequacy of utility infrastructure and the management practices that affect it; and
- establish a Web site to provide a central repository of information on comprehensive asset management so that drinking water and wastewater utilities have direct and easy access to information that will help them better manage their infrastructure.

AGENCY COMMENTS

GAO provided a draft of this report to EPA for review and comment. GAO received comments from officials within EPA's Office of Water and Office of the Chief Financial Officer, who generally agreed with the information presented in the report and GAO's recommendations. They further noted that while EPA has played a major role in bringing asset management practices to the water industry, significant additional activity could be undertaken, and they have placed a high priority on initiating activities similar to those suggested by GAO. The officials also made technical comments, which GAO incorporated as appropriate.

SECTION 1: INTRODUCTION

Drinking water and wastewater utilities are facing potentially significant investments over the next 20 years to upgrade an aging and deteriorated infrastructure, including underground pipelines, treatment, and storage facilities; meet new regulatory requirements; serve a growing population; and improve security. Adding to the problem is that many utilities have not been generating enough revenues from user charges and other local sources to cover their full cost of service. As a result, utilities have deferred maintenance and postponed needed capital improvements. To address these problems and help ensure that utilities can manage their needs cost-effectively, some water industry and government officials advocate the use of comprehensive asset management. Asset management is a systematic approach to managing capital assets in order to minimize costs over the useful life of the assets while maintaining adequate service to customers. While the approach is relatively new to the U.S. water industry, it has been used by water utilities in other countries for as long as 10 years.

Each year, the federal government makes available billions of dollars to help local communities finance drinking water and wastewater infrastructure projects. Concerns about the condition of existing infrastructure have prompted calls to increase financial assistance and, at the same time, ensure that the federal government's investment is protected. In recent years the Congress has been considering a number of proposals that would promote the use of comprehensive asset management by requiring utilities to develop and implement plans for maintaining, rehabilitating, and replacing capital assets, often as a condition of obtaining loans or other financial assistance.

The Federal Government Has Played a Major Role in Funding and Setting Requirements for Water Infrastructure

The federal government has had a significant impact on the nation's drinking water and wastewater infrastructure by (1) providing financial assistance to build new facilities and (2) establishing regulatory requirements that affect the technology, maintenance, and operation of utility infrastructure. As we reported in 2001, nine federal agencies made available about $46.6 billion for capital improvements at water utilities from fiscal years 1991 through 2000.[1] The Environmental Protection Agency (EPA) and the Department of Agriculture alone accounted for over 85 percent of the assistance, providing $26.4 billion and $13.3 billion, respectively, during the 10-year period; since then, the funding from these two agencies has totaled nearly $15 billion.[2] EPA's financial assistance is primarily in the form of grants to the states to capitalize the Drinking Water and Clean Water State Revolving Funds, which are used to finance improvements at local drinking water and wastewater treatment facilities, respectively.[3] As part of the Rural Community Advancement Program, Agriculture's Rural Utilities Service provides direct loans, loan guarantees, and grants to construct or improve drinking water, sanitary sewer, solid waste, and storm drainage facilities in rural communities.

In addition to its financial investment, EPA has promulgated regulations to implement the Safe Drinking Water Act and Clean Water Act, which have been key factors in shaping utilities' capital needs and management practices. For example, under the Safe Drinking Water Act, EPA has set standards for the quality of drinking water and identified effective technologies for treating contaminated water. Similarly, under the Clean Water Act, EPA has issued national minimum technology requirements for municipal wastewater utilities and criteria that states use to establish water quality standards that affect the level of pollutants that such utilities are permitted to discharge. Thus, the federal government has a major stake in protecting its existing investment in water infrastructure and ensuring that future investments go to utilities that are built and managed to meet key regulatory requirements.

Projected Drinking Water and Wastewater Infrastructure Needs Are Significant

Drinking water and wastewater utilities will need to invest hundreds of billions of dollars in their capital infrastructure over the next two decades, according to EPA; the Congressional Budget Office; and the Water Infrastructure Network, a consortium of industry, municipal, state, and nonprofit associations. As table 1 shows, the projected needs range from $485 billion to nearly $1.2 trillion. The estimates vary considerably, depending on assumptions about thc naturc of cxisting capital stock, replacement rates, and financing costs. Given the magnitude of the projected needs, it is important that utilities adopt a strategy to manage the repair and replacement of key assets as cost-effectively as possible and to plan to sustain their infrastructure over the long term.

Table 1. Recent Estimates of the Cost of Meeting Infrastructure Needs at Drinking Water and Wastewater Utilities

Dollars in billions

	Estimate						
		Capital investment only			Capital investment and financing		
Organization	Period covered	Drinking water	Wastewater	Total	Drinking water	Wastewater	Total
Congressional Budget Office[a]	2000-2019						
--Low					232	260	492
--High					402	418	820
EPA[b]	2000-2019						
--Low		154	331	485	178	402	580
--High		446	450	896	475	719	1,194
Water Infrastructure Network[c]	2000-2019	380	360	740	480	460	940

Source: GAO summary of infrastructure estimates from the Congressional Budget Office, EPA, and the Water Infrastructure Network.

Note: We did not assess the reliability of these data.

[a]See Congressional Budget Office, *Future Investment in Drinking Water and Wastewater Infrastructure,* (Washington, D.C.: November 2002). According to the report, the difference between the low and high estimates is attributable primarily to assumptions about the rate at which drinking water pipes are replaced, the savings associated with improved efficiency, the costs of controlling combined sewer overflows, and the length of the borrowing term. The estimates represent infrastructure costs as financed and thus include the estimated debt service paid from 2000 to 2019, whether for newly built projects or projects built before 2000.

[b]See U.S. Environmental Protection Agency, *The Clean Water and Drinking Water Infrastructure Gap Analysis*, EPA-816-R-02-020 (September 2002). According to the report, the difference between the low and high estimates is attributable to differing assumptions about infrastructure replacement rates.

[c]See Water Infrastructure Network, *Clean & Safe Water for the 21st Century* (April 2000). The estimates for capital investment and financing represent the capital costs associated with all investments during the 2000-2019 period as well as the interest paid over time on those investments. They differ from costs-as-financed estimates because they include debt service (principal and interest) paid after 2019 on investments during the two decades instead of debt service paid during that time on pre-2000 investments.

Local drinking water and wastewater utilities rely primarily on revenues from user rates to pay for infrastructure improvements. According to EPA's gap analysis, maintaining utility spending at current levels could result in a funding gap of up to $444 billion between projected infrastructure needs and available resources.[4] However, EPA also estimates that if utilities' infrastructure spending grows at a rate of 3 percent annually over and above inflation, the gap will narrow considerably and may even disappear. EPA's report concludes that utilities will need to use some combination of increased spending and innovative management practices to meet the projected needs.

The nation's largest utilities—those serving populations of at least 10,000— account for most of the projected infrastructure needs. For example, according to EPA data, large drinking water systems represent about 7 percent of the total number of community water systems, but account for about 65 percent of the estimated infrastructure needs. Similarly, about 29 percent of the wastewater treatment and collection systems are estimated to serve populations of 10,000 or more, and such systems account for approximately 89 percent of projected infrastructure needs for wastewater utilities. Most of the U.S. population is served by large drinking water and wastewater utilities; for example, systems serving at least 10,000 people provide drinking water to over 80 percent of the population.

Pipeline rehabilitation and replacement represents a significant portion of the projected infrastructure needs. According to the American Society of Civil Engineers, U.S. drinking water and wastewater utilities are responsible for an estimated 800,000 miles of water delivery pipelines and between 600,000 and 800,000 miles of sewer pipelines, respectively. According to the most recent EPA needs surveys, the investment needed for these pipelines from 1999 through 2019 could be as much as $137 billion.[5]

Several recent studies have raised concerns about the condition of the existing pipeline network. For example, in August 2002, we reported the results of a nationwide survey of large drinking water and wastewater utilities.[6] Based on the survey, more than one-third of the utilities had 20 percent or more of their pipelines nearing the end of their useful life; and for 1 in 10 utilities, 50 percent or more of their pipelines were nearing the end of their useful life. In 2001, a major water industry association predicted that drinking water utilities will face significant repair and replacement costs over the next three decades, given the average life estimates for different types of pipelines and the years since their original installation.[7] Other studies have made similar predictions for the pipelines owned by wastewater utilities.

Many Factors Have Contributed to Deteriorating Utility Infrastructure

EPA and water industry officials cite a variety of factors that have played a role in the deterioration of utility infrastructure; most of these factors are linked to the officials' belief that the level of ongoing investment in the infrastructure has not been sufficient to sustain it. For example, according to EPA's Assistant Administrator for Water, the pipelines and plants that make up the nation's water infrastructure are aging, and maintenance is too often deferred. He predicted that consumers will face sharply rising costs to repair and replace the infrastructure. Similarly, as the Water Environment Research Foundation reported in 2000, "years of reactive maintenance and minimal expenditures on sewers have left a huge backlog of repair and renewal work."[8]

Our nationwide survey of large drinking water and wastewater utilities identified problems with the level of revenues generated from user rates and decisions on investing these revenues.[9] For example:

- Many drinking water and wastewater utilities do not cover the full cost of service—including needed capital investments and operation and maintenance costs—through their user charges. Specifically, a significant percentage of the utilities serving populations of 10,000 or more—29 percent of the drinking water utilities and 41 percent of the wastewater utilities—were not generating enough revenue from user charges and other local sources to cover their costs.
- Many drinking water and wastewater utilities defer maintenance and needed capital improvements because of insufficient funding. About one-third of the utilities deferred maintenance expenditures in their most recent fiscal year;[10] similar percentages of utilities reported deferring minor capital improvements and major capital improvements. About 20 percent of the utilities had deferred expenditures in all three categories.
- For many utilities, a significant disparity exists between the actual rehabilitation and replacement of their pipelines and the rate at which utility managers believe rehabilitation and replacement should occur. We found that only about 40 percent of the drinking water utilities and 35 percent of the wastewater utilities met or exceeded their desired rate of pipeline rehabilitation and replacement. The remaining utilities did not meet their desired rates. Roughly half of the utilities

actually rehabilitated or replaced 1 percent or less of their pipelines annually.

Utility managers also lack the information they need to manage their existing capital assets. According to our survey, many drinking water and wastewater utilities either do not have plans for managing their assets or have plans that may not be adequate in scope or content. Specifically, nearly one-third of the utilities did not have plans for managing their existing capital assets. Moreover, for the utilities that did have such plans, the plans in many instances did not cover all assets or did not contain one or more key elements, such as an inventory of assets, assessment criteria, information on the assets' condition, and the planned and actual expenditures to maintain the assets.[11]

Comprehensive Asset Management Focuses on Efficiently Managing Capital Assets

Comprehensive asset management has gained increasing recognition within the water industry as an approach that could give utilities the information and analytical tools they need to manage existing assets more effectively and plan for future needs. Using asset management concepts, utilities and other organizations responsible for managing capital infrastructure can minimize the total cost of designing, acquiring, operating, maintaining, replacing, and disposing of capital assets over their useful lives, while achieving desired service levels. Figure 1 shows some of the basic elements of comprehensive asset management and how the elements build on and complement each other to form an integrated management system.

Experts within and outside the water industry have published manuals and handbooks on asset management practices and how to apply them.[12] While the specific terminology differs, some fundamental elements of implementing asset management appear consistently in the literature.

- *Collecting and organizing detailed information on assets.* Collecting basic information about capital assets helps managers identify their infrastructure needs and make informed decisions about the assets. An inventory of an organization's existing assets generally should include (1) descriptive information about the assets, including their age, size, construction materials, location, and installation date; (2) an assessment of the assets' condition, along with key information on operating, maintenance, and repair history, and the assets' expected and remaining useful life; and (3) information on the assets' value, including historical cost, depreciated value, and replacement cost.

- *Analyzing data to set priorities and make better decisions about assets.* Under asset management, managers apply analytical techniques to identify significant patterns or trends in the data they have collected on capital assets; help assess risks and set priorities; and optimize decisions on maintenance, repair, and replacement of the assets. For example:
- *Life-cycle cost analysis.* Managers analyze life-cycle costs to decide which assets to buy, considering total costs over an asset's life, not just the initial purchase price. Thus, when evaluating investment alternatives, managers also consider differences in installation cost, operating efficiency, frequency of maintenance and repairs, and other factors to get a cradle-to-grave picture of asset costs.
- *Risk/criticality assessment.* Managers use risk assessment to determine how critical the assets are to their operations, considering both the likelihood that an asset will fail and the consequences—in terms of costs and impact on the organization's desired level of service—if the asset does fail.[13] Based on this analysis, managers set priorities and target their resources accordingly.
- *Integrating data and decision making across the organization.* Managers ensure that the information collected within an organization is consistent and organized so that it is accessible to the people who need it. Among other things, the organization's databases should be fully integrated; for instance, financial and engineering data should be compatible, and ideally each asset should have a unique identifier that is used throughout the organization. Regarding decision making, all appropriate units within an organization should participate in key decisions, which ensures that all relevant information gets considered and encourages managers to take an organizationwide view when setting goals and priorities.
- *Linking strategy for addressing infrastructure needs to service goals, operating budgets, and capital improvement plans.* An organization's goals for its desired level of service—in terms of product quality standards, frequency of service disruptions, customer response time, or other measures—are a major consideration in the organization's strategy for managing its assets. As managers identify and rank their infrastructure needs, they determine the types and amount of

investments needed to meet the service goals. Decisions on asset maintenance, rehabilitation, and replacement are, in turn, linked to the organization's short- and long-term financial needs and are reflected in the operating budget and capital improvement plan, as appropriate.

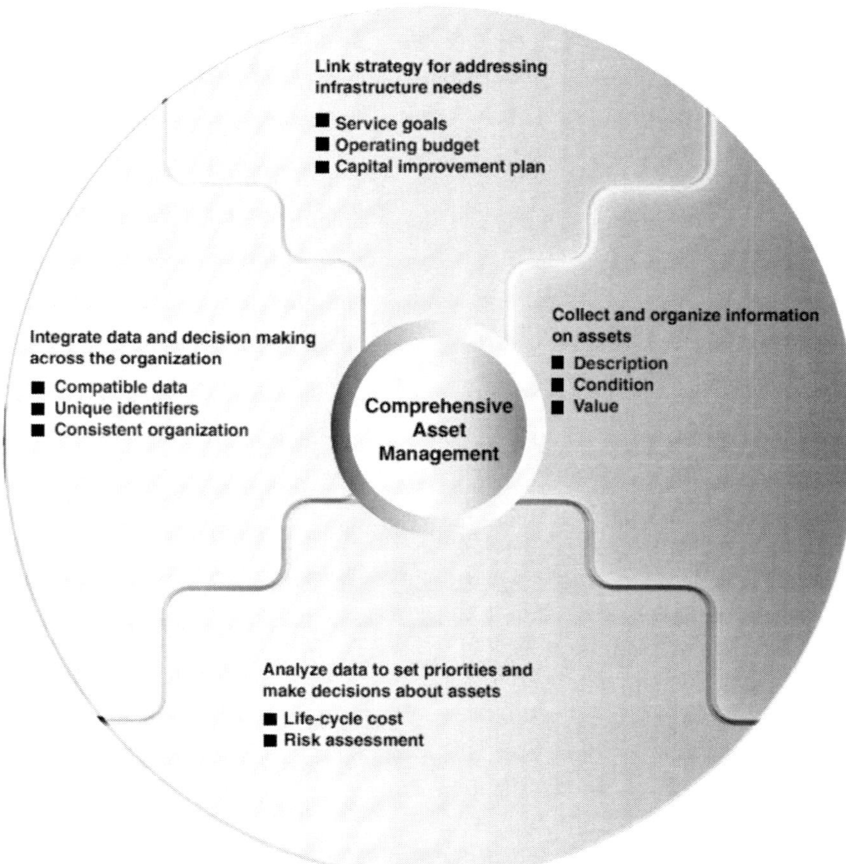

Source: GAO.

Figure 1. Elements of Comprehensive Asset Management.

Implementing the basic elements of asset management is an iterative process that individual organizations may begin at different points. Within the water industry, for example, some utilities may start out by identifying their infrastructure needs, while other utilities may take their first step by setting goals for the level of service they want to provide. The interrelationship between the elements of asset management can alter an organization's strategy for managing its assets. For example, once an organization has completed a risk assessment, it may scale back its efforts to compile a detailed inventory of assets to focus initially on those assets determined to be critical. Similarly, as information on infrastructure needs and priorities improves, managers reexamine the level of planned investments, considering the impact on both revenue requirements and the level of service that can be achieved. According to advocates of asset management, while many organizations are implementing certain aspects of the process, such as maintaining an inventory of assets and tracking maintenance, these organizations are not realizing the full potential of *comprehensive* asset management unless all of the basic elements work together as an integrated management system.

Strategy for Adopting and Progress toward Implementing Comprehensive Asset Management Varies

As the description of asset management indicates, implementing this approach is not a step-by-step, linear process. Asset management is an integrated system that utilities and other organizations can implement in a number of different ways, depending on what makes sense for their particular organization. In the United States, some drinking water and wastewater utilities, for example, are taking a more strategic approach, initially investing their resources in planning for asset management. Other utilities are focusing initially on collecting data. Another variation is that some utilities are adopting asset management on a utilitywide basis, while others are piloting the approach at a single facility or department or are targeting critical assets utilitywide. The level of sophistication with which asset management concepts are applied within a utility can also vary, depending on the size and complexity of the operations and the resources that the utility can devote to implementation.

Comprehensive asset management is a relatively new concept for drinking water and wastewater utilities in the United States. According to EPA and major water industry organizations, few utilities are implementing comprehensive asset management, and those that have done so are almost exclusively larger entities. In addition, for the most part, the domestic utilities that have adopted asset management are in the early stages of implementation. Few utilities have been involved in the process for longer than 2 to 3 years.

Although relatively new to the U.S. water industry, comprehensive asset management has been used for about 10 years by water utilities in Australia and New Zealand, where the national governments have strongly endorsed the concept. In each case, the driving force behind the use of asset management was legislation that called for water utilities to improve their financial management. In Australia, the law requires utilities to recover the full cost of service, while in New Zealand the law requires utilities to depreciate their assets annually and use cost-benefit analysis, among other things. The national governments of Australia and New Zealand each published guidebooks on asset management, and engineering groups in the two countries jointly developed a comprehensive manual on managing infrastructure assets.[14]

Asset management is seen as a means of improving utility infrastructure elsewhere in the world. For example, in the United Kingdom, utilities must develop asset management plans that identify the level of investment required to maintain and improve capital assets every 5 years; annual audits help ensure that planned improvements are made. Similarly, in 2002, the legislature in Ontario, Canada enacted a law requiring municipalities to develop plans for recovering the full cost of service to ensure that drinking water and wastewater systems are adequately funded.

Objectives, Scope, and Methodology

The Ranking Minority Member, Senate Committee on Environment and Public Works, asked us to examine the use of comprehensive asset management at drinking water and wastewater utilities in the United States. This report examines (1) the potential benefits of asset management for water utilities and the challenges that could hinder its implementation and (2) the role that the federal government might play in encouraging utilities to implement comprehensive asset management.

To conduct our work, we reviewed relevant studies, handbooks, training materials, and other documents related to comprehensive asset management and its implementation, particularly for managing the infrastructure at drinking water and wastewater utilities. At the federal level, we obtained information from EPA's Office of Ground Water and Drinking Water and Office of Wastewater Management, the offices that, along with the states, are responsible for overseeing drinking water and wastewater utilities. We also obtained information on other federal agencies with experience in asset management, predominantly the Federal Highway Administration in the U.S.

Department of Transportation, and financial standards promulgated by the Governmental Accounting Standards Board. For site-specific information, our review included over 50 individual utilities from the United States, Australia, and New Zealand— including 15 U.S. utilities at which we conducted structured interviews.

Other sources of information included the following:

- state associations, including the Association of State Drinking Water Administrators and the Association of State and Interstate Water Pollution Control Administrators;
- major industry groups, including the American Public Works Association, American Water Works Association, Association of Metropolitan Sewerage Agencies, Association of Metropolitan Water Agencies, National Association of Water Companies, National Rural Water Association, Water Environment Federation, and Water Services Association of Australia;
- engineering and consulting firms with experience in helping utilities implement asset management, including Brown and Caldwell; CH2M Hill; Metcalf and Eddy, Inc.; Municipal and Financial Services Group; PA Consulting Group; and Parsons Corporation in the U.S.; GHD Pty. Ltd. in Australia; and Meritec in New Zealand;
- several state and regional regulatory agencies in Australia and New Zealand; and
- EPA-funded state and university-based training and technical assistance centers.

To obtain information on the benefits and challenges of asset management, we conducted initial interviews with 46 domestic drinking water and wastewater utilities that knowledgeable government and water industry officials identified as implementing comprehensive asset management. To obtain more detailed information, we conducted structured interviews with officials from 15 of the 46 utilities. We selected the 15 utilities based on two criteria: (1) they reported or anticipated achieving quantitative benefits from asset management or (2) they represented smaller entities.[15] (See app. I for a list of the 15 utilities we selected for structured interviews.) In total, 12 of the 15 utilities were relatively large, serving populations ranging from 300,000 to 2,500,000; the remaining three were significantly smaller, serving populations ranging from 3,000 to 67,100. Because of the small number of utilities that we interviewed in depth and the way in which they were selected, our results are not generalizable to the larger universe of domestic drinking water and wastewater utilities.

Because of the utilities' limited experience in implementing asset management, we supplemented the information obtained from domestic utilities with information from six utilities and five government agencies in Australia and New Zealand, two countries that have taken the lead in implementing comprehensive asset management. (See app. II for a list of the utilities and government agencies we contacted in Australia and New Zealand.) Outside the water industry, we consulted with the Private Sector Council, which identified two companies—The Gillette Company and SBC Communications, Inc.—with long-standing experience in using comprehensive asset management in their respective fields. We interviewed officials from these companies to obtain their perspectives on the benefits and challenges of implementing asset management.

For information on the potential federal role in promoting asset management at water utilities, we obtained information from EPA's Office of the Chief Financial Officer, Office of Ground Water and Drinking Water, and Office of Wastewater Management on the activities that EPA is currently sponsoring, including the development of informational materials on asset management; activities by EPA-funded, state and university-based training and technical assistance centers; and various studies and research projects. We also discussed options for a federal role in promoting asset management with officials from water industry associations, EPA, and the 15 utilities selected for structured interviews. In addition, with the help of organizations and officials experienced in asset management, we identified the U.S. Department of Transportation as being at the forefront of federal involvement in this issue. We obtained and reviewed information about the department's initiatives from the Office of Asset Management within the Federal Highway Administration.

We conducted our work between March 2003 and March 2004 in accordance with generally accepted government auditing standards.

Comments from the Environmental Protection Agency

We provided a draft of this report to EPA for review and comment. We received comments from officials within EPA's Office of Water and Office of the Chief Financial Officer, who generally agreed with the information presented in the report and our recommendations. They further noted that while EPA has played a major role in bringing asset management practices to the water industry, significant additional activity could be undertaken, and they have placed a high priority on initiating activities similar to those we suggested. The officials also made technical comments, which we incorporated as appropriate.

SECTION 2: WATER INDUSTRY OFFICIALS REPORT MANY BENEFITS FROM ASSET MANAGEMENT DESPITE IMPLEMENTATION CHALLENGES

While comprehensive asset management is relatively new to most drinking water and wastewater utilities in the United States, some utilities say they have already benefited from this approach and have also encountered certain challenges. The utilities reported benefiting from (1) improved decision making because they have better information about their capital assets and (2) improved relationships with governing authorities, ratepayers, and other stakeholders because they are better able to communicate information on infrastructure needs and improvement plans. While water industry officials identified benefits associated with comprehensive asset management, we found that reported savings should be viewed with caution.

Among the challenges of implementing asset management, utility officials cited the difficulty of (1) collecting the appropriate data and managing it efficiently and (2) making the cultural changes necessary to integrate information and decision making across departments. In addition, the officials reported that the short-term budget and election cycles typical of utility governing bodies make it difficult to meet the long-term capital investment planning needs of asset management. Although smaller utilities face more obstacles to implementing asset management than larger utilities, principally because of limited resources, they can also benefit from applying asset management concepts.

Utilities Cite Many Benefits from Asset Management and Some Cautions About Reported Savings

U.S. utilities expect to reap significant benefits from the data they collect, analyze, and share through an asset management approach. With these data, utilities expect to make more informed decisions on maintaining, rehabilitating, and replacing their assets, thereby making their operations more efficient. Utilities can also use these data to better communicate with their governing bodies and the public, which should help them to make a sound case when seeking rate increases. Although water industry officials identified financial and other benefits from using asset management, reported savings should be viewed with caution because, for instance, comprehensive asset management may be implemented concurrently with other changes in management practices or operational savings may be offset by increases in capital expenditures.

Comprehensive Asset Management Can Improve Decisions about Maintaining, Rehabilitating, and Replacing Capital Assets

Collecting, sharing, and analyzing data through comprehensive asset management can help utilities to make more informed decisions about maintaining, rehabilitating, and replacing their assets. In particular, utilities can use the information collected and analyzed to prevent problems and allocate their maintenance resources more effectively. For example:

- Better information enabled the Massachusetts Water Resources Authority to improve its maintenance decisions and eliminate some unneeded maintenance activities.[1] For example, in an effort to optimize maintenance practices in one of their treatment plants, utility officials reassessed maintenance practices for 12 equipment systems, such as different types of pumps. By using the assessment results to improve maintenance planning for these assets, the utility decreased the labor hours spent on preventive maintenance by 25 percent from the hours recommended by the original equipment manufacturers, according to utility officials. Similarly, in analyzing its maintenance practices, the Massachusetts Water Resources Authority found it was lubricating some equipment more often than necessary. By decreasing the frequency of oil changes, the utility reported it saved approximately $20,000 in oil purchase and disposal costs. In addition, the utility extended the life of its assets by decreasing the lubrication—over-lubrication can cause equipment parts to fail prematurely.
- Seattle Public Utilities used asset management to better target its maintenance resources.[2] As part of the utility's asset management strategy, officials used a risk management approach, calculating the likelihood and impact of a rupture for the utility's sewer and drainage

pipes. To determine the likelihood of rupture, officials considered such factors as a pipe's age, material, and proximity to a historical landfill or steep slope. To determine the impact of a rupture, they examined factors such as a pipe's size, location, and historical cost of repair. As a result of this analysis, utility officials identified 15 percent of their pipes as high risk, or "critical"—such as larger, older pipes located beneath downtown Seattle. They shifted resources to maintain and rehabilitate these pipes. The officials considered the remaining 85 percent of pipes as noncritical, or, lower risk, because their failure was less likely or because a breakage would affect a limited number of customers, be repaired relatively quickly, and require minimal resources. For these pipes, the utility decided not to perform any preventive maintenance activities, only making repairs as needed. By taking this approach, utility officials believe they are using their staff resources more efficiently and that, over time, they will reduce their maintenance costs.

Comprehensive asset management also helps managers to make more informed decisions about whether to rehabilitate or replace assets, and once they decide on replacement, to make better capital investment decisions. For example:

- According to utility managers at the Louisville Water Company, the utility developed its Pipe Evaluation Model in the early 1990s as a tool for ranking its 3,300 miles of aging pipes and water mains for rehabilitation and replacement.[3] The pipe program includes many of the key principles and practices of comprehensive asset management: for instance, it integrated data about the age of the pipes with data about their maintenance history. In analyzing this information, managers discovered that two vintages of pipes—those built between 1862 and 1865 and between 1926 and 1931—had the highest number of breaks per 100 miles of pipeline. Consequently, they decided to replace the pipes from those two periods. The model also showed that pipes installed between 1866 and 1925 were fairly reliable, thus these pipes were targeted for rehabilitation rather than replacement. The utility is lining the interior of these pipes with cement, which is expected to extend their life by about 40 years. Furthermore, utility managers told us that their pipe model and other practices that use asset management principles have helped reduce the frequency of water main breaks from 26 to 22.7 per hundred miles and the frequency of leaks from joints from 8.2 to 5.6 per hundred miles.

- In implementing its asset management approach, managers at the Sacramento Regional County Sanitation District reassessed a proposed investment in new wastewater treatment tanks and decided on a less expensive option, thereby saving the utility approximately $12 million.[4] During this reassessment, managers found that increasing preventive maintenance on existing tanks would lower the risk of shutdown more cost-effectively than adding a new set of tanks. Utility officials commented that their implementation of asset management helped change their decision-making process by, among other things, bringing together staff from different departments to ensure more complete information, and more effectively using the data to understand investment options.
- As a part of its asset management strategy, Seattle Public Utilities established an asset management committee, comprised of senior management from various departments, to ensure appropriate decision making about the utility's capital improvement projects. For every capital improvement project with an expected cost over $250,000, project managers must submit a plan to the committee that (1) defines the problem to be solved, (2) examines project alternatives, (3) estimates the life-cycle costs of the alternatives, (4) analyzes the possible risks associated with the project, and (5) recommends an alternative. According to utility officials, implementing this process has led to deferring, eliminating, or altering several capital improvement projects, and contributing to a reduction in the utility's 2004 capital improvement project budget for water of more than 8 percent. For instance, after drafting new water pressure standards, the utility eliminated the need for some new water mains. It developed an alternative plan to provide more localized solutions to increase water pressure, resulting in expected savings of $3 million. In another case, the utility reassessed alternatives to replacing a sewer line located on a deteriorating trestle, ultimately opting to restore and maintain the existing wood trestle and make spot repairs to the sewer line, which resulted in an estimated savings of $1.3 million.

Finally, comprehensive asset management helps utilities share information across departments and coordinate planning and decision making. In this way, utility managers can reduce duplication of efforts and improve the allocation of staff time and other resources. For example, managers at Eastern Municipal Water District used asset management to improve their business practices, which they saw as compartmentalized and inefficient.[5] In one instance, they examined their decentralized maintenance activities. The utility had two maintenance crews who worked throughout the system, in different shifts and reported to managers at four different facilities. In addition, the utility's work order system was inefficient; for example, when different crew members independently reported the same maintenance need, managers did not notice the duplication because the problem was described in different terms (e.g., as a "breaker failure" by one crew member and as a "pump failure" by another). Finally, in some instances, work crews would arrive at a site only to find that needed maintenance work had already been completed. To improve the system, utility officials (1) centralized maintenance by making one person responsible for scrutinizing and setting priorities for all work orders and (2) established a standardized classification of assets, which helped maintenance staff use the same terminology when preparing work orders. Utility officials report that taking these steps allowed them to identify and eliminate work orders that were unnecessary, already completed, or duplicates, which ultimately reduced their maintenance work backlog by 50 percent.

The private sector companies we visited agreed that using a comprehensive asset management approach improved their decision making. Specifically, by improving their data, analyzing these data, and centralizing management decision making, managers at SBC Communications, Inc., reported that they have made better capital investment decisions and allocated resources more efficiently. Managers at The Gillette Company reported that they consider life-cycle costs and other factors to assess investment alternatives and, ultimately, make better investment decisions.

Comprehensive Asset Management Can Help Utilities Justify Rate Increases and Proposed Projects to Their Customers and Governing Bodies

The utilities we contacted reported that comprehensive asset management also benefits their relations with external stakeholders by (1) making a sound case for rate increases to local governing bodies and ratepayers; (2) improving their bond rating with credit rating agencies, and (3) better demonstrating compliance with federal and state regulations.

Making a Sound Case for Rate Increases

Some utilities have used, or expect to use, the information collected through comprehensive asset management to persuade elected officials to invest in drinking water and wastewater infrastructure through rate increases. For example, the Louisville Water Company reported that in the early 1990s it used the asset information it had gathered and analyzed to convince its local governing board that its current rates would not cover its expected costs and that the utility needed a rate increase to cover its anticipated rehabilitation and replacement needs. The board approved a set-aside of $600,000 for an infrastructure rehabilitation and replacement fund as a part of the requested rate increase in 1993, and, according to one utility official, has been supportive of including funds for asset rehabilitation and replacement as a part of rate requests since then. Furthermore, the utility manager requested that the amount of the set-aside gradually increase to $3 million over the next 5 years. According to this official, the board not only approved this request, it also increased the rates to support the fund sooner than the utility manager had requested. According to several other utilities that have begun to implement comprehensive asset management, this approach should enable them to justify needed rate increases from their governing bodies. Similarly, Australian and New Zealand officials we interviewed stated that the data from asset management helps utilities make a more credible case for rate increases from their governing bodies.

Utility managers can also use the information they provide to their governing boards as a basis for evaluating and deciding on trade-offs between service levels and rates. For example, according to an official at South Australian Water Corporation, using asset management practices, he was able to suggest a range of funding alternatives to the utility's governing body.[6] The utility managers conducted statistical modeling on the asset information they collected (e.g., pipe performance history and financial information) and, using this analysis, predicted the approximate number of pipe breaks at various levels of funding. Understanding the trade-offs between lower rates and higher numbers of pipe breaks, the governing body could make an informed decision about what the appropriate level of service was for their community.

Improving the Bond Rating

Comprehensive asset management also has the potential to improve a utility's bond rating, a benefit that translates into savings through lower interest rates on loans and bonds. When deciding on a utility's bond rating, credit rating agencies consider criteria related to comprehensive asset management, such as the utility's management strategies and its planning for asset replacement. For example, according to a representative from one credit rating agency, asset management shows that a utility is considering future costs. He would therefore expect a utility with an asset management plan that looks at future capital and operating costs and revenues to receive a higher bond rating than a utility that does not sufficiently consider those future needs, even if that utility has a better economy and a higher tax base.

Some local officials believe that comprehensive asset management played a role in the bond ratings they received, or will do so in the future. For example, the finance director of the small northeastern city of Saco, Maine, told us that she believes that the city's decision to use asset management practices—such as maintaining an up-to-date asset inventory, periodically assessing the condition of the assets, and estimating the funds necessary to maintain the assets at an acceptable level each year—contributed to the credit rating agencies' decision to increase the city's bond rating, which resulted in an expected savings of $2 million over a 20-year period.[7] Similarly, a utility official at Louisville Water Company told us that asset management practices, such as strategically planning for the rehabilitation and replacement of its aging assets, helps the utility maintain its strong bond rating.

Better Demonstrating Compliance with Federal and State Regulations

According to several utility managers we interviewed, comprehensive asset management can be used to help comply with regulations. For example:

- Comprehensive asset management practices played a role in improving their utility's compliance with existing regulations. Specifically, among other things, asset management practices such as identifying and maintaining key assets led to fewer violations of pollutant discharge limitations under the Clean Water Act. At Western Carolina Regional Sewer Authority, for instance, the number of these violations decreased from 327 in 1998 (about the time that the utility began implementing asset management) to 32 violations in 2003.[8]
- At the Charleston Commissioners of Public Works,[9] utility officials told us that if they had not had asset management in place it would be difficult to meet the rehabilitation program and maintenance program elements of EPA's draft capacity, management, operation, and maintenance regulations for wastewater utilities.[10] For instance, the draft regulations would require that wastewater utilities identify and implement rehabilitation actions to address structural deficiencies.

Because the utility has implemented asset management practices, such as assessing the condition of its pipes and identifying those most in need of rehabilitation, it can better target its resources to rehabilitate pipes in the worst condition, and, in the process, meet the proposed standards for rehabilitation.

Although Water Industry Officials Identified Financial and Other Benefits from Asset Management, Reported Savings Should Be Viewed with Caution

Many of the U.S. utilities we interviewed were still in the early stages of implementing asset management and most had not measured financial savings. However, many water industry officials expect asset management to result in overall cost savings. Specifically, several officials told us they expect that asset management will slow the rate of growth of utilities' capital, operations, and maintenance costs over the coming years. Nevertheless, total costs will rise because of the need to replace and rehabilitate aging infrastructure.

At least one U.S. utility has estimated the overall savings it will achieve using comprehensive asset management. Specifically, an engineering firm projected that asset management would reduce life-cycle costs for the Orange County Sanitation District by about $350 million over a 25-year period.[11] Among other data, the engineering firm used the utility's available operating expenditure information (operations, maintenance, administration, and depreciation data) and capital improvement program expenditures (growth/capacity, renewal/replacement, and level of support data) to model the projected life-cycle cost savings.

Additionally, some of the Australian utilities we interviewed reported financial savings. For example, officials at Hunter Water Corporation reported significant savings in real terms between fiscal years 1990 and 2001: a 37 percent reduction in operating costs;[12] improved service standards for customers, as measured by such factors as water quality and the number of sewer overflows; and a reduction of more than 30 percent in water rates for customers.[13] Hunter Water officials believe that they achieved these efficiencies as a result of asset management.

Though utility officials have made some attempts to quantify the impact of asset management, they also cited reasons for exercising caution in interpreting reported savings and other benefits. First, benefits such as operating cost reductions should not be considered in isolation of other utility costs. A utility cannot consider reductions in operating costs a net benefit if, for instance, savings in operational costs are offset by an increase in the utility's capital expenditures. Furthermore, reductions in operating costs may be caused by increases in capital expenditures because, for example, newer assets may require less maintenance and fewer repairs. In the case of the Hunter Water Corporation, the utility's capital expenditures were at about the same level in 2001 as in 1991, despite some fluctuation over the period.[14]

Second, other factors might have contributed to financial and other benefits. For example, a utility may be implementing other management initiatives concurrently with asset management and may not be able to distinguish the benefits of the various initiatives. In addition to using an asset management approach, for instance, some U.S. utilities we interviewed used an environmental management system, which shares some of the same components as asset management.[15] Some of these utilities told us that they could not separate the benefits of asset management from those achieved as a result of their environmental management systems.

In addition, reported savings from asset management can be misleading without complete information on how the savings estimates are derived. For example, a widely distributed graph shows an estimated 15 percent to 40 percent savings in life-cycle costs for 15 wastewater utilities in Australia. EPA and others used the graph as a basis for projecting savings for U.S. utilities. However, the graph was mislabeled at some point—the reported reductions in life-cycle costs were actually reductions in operating costs.[16] As we have already noted, operating costs reductions alone do not provide enough information to determine the net benefit of implementing asset management.

Utilities Face Challenges in Successfully Implementing Comprehensive Asset Management

Despite the acknowledged benefits of comprehensive asset management, utilities face three key challenges that may make implementing this approach difficult. First, to determine the condition of current assets and the need for future investment, utilities have to gather and integrate complete and accurate data, which may require significant resources. Second, successful implementation requires cultural change—departments long accustomed to working independently must be willing to coordinate and share information. Finally, utilities may find that their efforts to focus on long-term planning conflict with the short-term priorities of their governing bodies. These three challenges may be more difficult for smaller utilities because they have fewer financial, staff, and technical resources.

Asset Management Requires Utilities to Collect Complete and Accurate Data

The difficulties utilities experience gathering data to implement asset management depend on the (1) condition of their existing data, (2) ability to coordinate existing data across departments, (3) need to upgrade technology, and (4) ability to sustain complete and accurate data. One industry official noted that larger utilities, in particular, may have a more difficult time gathering and coordinating data because they typically possess a substantial number of assets. Nevertheless, utility officials and water association representatives agree that utilities should not allow these data challenges to prevent them from implementing asset management. These officials emphasized that utilities should begin implementing asset management by using the data they already possess, continuing data collection as they perform their routine repair and maintenance activities, or focusing data collection efforts on their most critical assets.

Existing Data May Be Incomplete and Inaccurate

Domestic and international water officials emphasize the importance of obtaining, integrating, and sustaining good data for decision making. This is no small challenge. According to the Association of Metropolitan Sewerage Agencies and the *International Infrastructure Management Manual*, utilities generally need the following types of data to begin implementing asset management:

- age, condition, and location of the assets;
- asset size and/or capacity;
- valuation data (e.g., original and replacement cost);
- installation date and expected service life;
- maintenance and performance history; and
- construction materials and recommended maintenance practices.

According to utility officials and industry handbooks, utilities sometimes have incomplete or inaccurate historical data about their assets. For example:

- An official at the Augusta County Service Authority noted that the utility did not possess a great deal of detailed historical data about its assets.[17] For example, its asset ledger would indicate that "a pump station was installed at a particular location in 1967," but would not provide any additional information about the assets, such as the individual components that make up this system. Similarly, the official told us that the utility's prior billing system did not maintain historical data about its customers' water usage rates. As a result, the management team found it difficult to adequately forecast their needed rate increases because they lacked historical information about water consumption.

- According to an East Bay Municipal Utility District official, the utility lacked detailed maintenance data on its assets before 1990 because maintenance workers had not consistently reported repairs to a central office.[18]

Given these problems, utility managers may have to invest a significant amount of time and resources to gather necessary data, particularly data about the condition of their thousands of miles of buried pipelines. Understandably, utilities are unwilling to dig up their pipelines to gather missing data. However, utilities may be able to derive some information about the condition of these pipes to the extent they have information on the pipes' age, construction material, and maintenance history. In addition, utilities may choose to align their data collection with their ongoing maintenance and replacement activities. These approaches, however, may require new technology, which may mean a financial investment. For example:

- Tacoma Water equipped its staff with laptop computers, which allows them to access their geographic information system—software that can track where assets are located—while they are in the field.[19] As the staff perform their routine repair and rehabilitation activities, they can record and update data about an asset's condition, performance, and maintenance history.
- Similarly, the Department of Public Works in Billerica, Massachusetts, provided its field staff with handheld electronic devices programmed with a simple data collection template, which allows its staff to more accurately record information about its assets and their condition.[20] Consequently, the field staff can enter more accurate information about the utility's assets into its central asset inventory.

Utilities also reported difficulty collecting and applying information about the manufacturer's recommended techniques for optimizing their maintenance practices for their assets. Since no central clearinghouse of information on optimal maintenance practices is readily available, these utilities have had to invest their own time and resources to develop this information. For example:

- According to an official at Des Moines Water Works, the utility discovered that the manufacturer's recommended maintenance practices often conflicted with the utility's experience with the same asset.[21] This official pointed out that the manufacturer's estimate for maintenance was always higher than the utility's experience. Given these inconsistencies, the official noted, all utilities would benefit from the development of a central industry clearinghouse that provided information about the recommended maintenance practices for certain assets.
- Similarly, an official at East Bay Municipal Utility District noted a significant difference between the manufacturer's recommended maintenance practices and the utility's experience with optimized maintenance. As a result, the utility has invested a significant amount of time in developing optimal maintenance practices for its assets and minimizing the risk of asset failure.

While utilities need complete and accurate data for decision making, they also need to balance data collection with data management. Utilities may fall prey to data overload—collecting more data than they have the capacity to manage. For example, according to an official at the Augusta County Service Authority, while the utility has collected extensive infrastructure data, it has not invested enough of its resources into making these data useful for decision making. This official told us that utilities need to develop a data management strategy that identifies the types of data they need and the uses of these data for decision making. Without such a strategy, utilities gathering data will reach a point of diminishing returns. According to an official at the National Asset Management Steering Group in New Zealand, utilities should begin to implement asset management by identifying their critical assets and targeting their data-gathering activities toward the critical information they need in order to make decisions about these assets. An official also recommended that utilities begin implementation by using their existing data—even though the data may not be completely accurate—and refine this information as they improve and standardize their data collection processes.

Coordinating Data Across Departments May Be Difficult

According to utility officials, coordinating data can be difficult because the data come from several different departments and from different sources within the departments. Furthermore, one industry handbook notes that a utility's departments typically maintain different types of data about the same assets, which are formatted and categorized to meet each department's individual needs and objectives.[22] For example, the finance department may record an asset's size in terms of square footage, while the engineering department may define an asset's size in terms of pipeline diameter. Utilities adopting asset management need to coordinate these data to develop a central asset inventory. Table 2 shows the typical sources of data for a central inventory.

Table 2. Typical Sources of Data for a Central Asset Inventory

Department maintaining data	Source of data	Types of data that may be available
Operations and Maintenance	Operations and maintenance manuals	Location, size, manufacturer, and materials of construction
	Maintenance management system	Location, size, manufacturer, materials of construction, performance history, maintenance history, and original cost
	Records of original asset drawings	Location, size, manufacturer, and materials of construction
Engineering	Geographic information system	Location, size, and materials of construction
	Records of original asset drawings	Location, size, manufacturer, and materials of construction
Finance	Fixed asset inventory	Size (e.g., square footage or acreage), installation date, and estimated value

Source: GAO.

Note: Summary of material from two industry handbooks, *Managing Public Infrastructure Assets to Minimize Cost and Maximize Performance* and the *International Infrastructure Management Manual*.

Utility managers told us it was challenging to develop a standard data format for their central asset inventories. For example:

- As previously noted, Eastern Municipal Water District's work order system was inefficient because crew members from different facilities did not use the same terms in describing maintenance problems. To eliminate these inefficiencies, the utility invested a great deal of time and resources to standardize its terms and asset classification and implement a computerized maintenance management system.
- According to a Louisville Water Company official, improving and validating the utility's data was a challenge. Over the years, the utility has acquired between 12 and 20 smaller utilities. Each of these smaller utilities maintained its own asset data, which were not always reliable or maintained in the same format. The utility invested a great deal of time to validate these data and coordinate them into its central asset inventory.
- Similarly, according to an official at the South Australian Water Corporation, developing a central asset inventory was particularly difficult because each of the utility's departments used different terms to refer to the same asset. The utility refined its data collection practices by training its employees on how to record data in a standard format.

Utilities May Need to Upgrade Their Technology

The utility officials we spoke to also had to address problems in coordinating data maintained in different and incompatible software programs. A Water Environment Research Foundation survey of utility managers, regulators, and industry consultants cited developing an asset information management system that meets the needs of all users as the most difficult element of asset management to implement. Without an integrated information management system, utilities found it difficult to develop data for decision making, and they found that they had to invest time and money to enter these data into a central database. For example:

- According to a Greater Cincinnati Water Works official, the utility wanted to integrate information about its assets' location and maintenance history to efficiently dispatch staff to repair sites.[23] However, the data for this report were stored in two separate and incompatible computer systems. To produce this information, the utility needed to re-enter the relevant data from each of these systems into a central asset database.

- Similarly, an official at Melbourne Water Corporation said that as his utility began to adopt asset management, it realized that it maintained relevant data in different computer systems, such as its computerized maintenance management system and its geographic information system.[24] To address this fragmentation, the utility had to assign staff to consolidating its data into a central database to allow for easy integration.

As utilities coordinate their data systems, they may need to upgrade their existing technology, which can represent a significant financial investment.

For example, Augusta County Service Authority has requested $100,000 to purchase data integration software, which would allow it to coordinate information from several different computer systems. However, as of September 2003, this request had not been approved, in part because the software may not directly affect the utility's profits or improve its service, making the governing body reluctant to finance the purchase. Similarly, St. Paul Regional Water Services recognized that it would need to purchase a geographic information system as the basis for integrating all departments' data.[25] However, the official noted that the utility could not purchase this system for another 4 years because it would cost several million dollars to purchase the system, enter data, and train its staff to operate the new system.

Utilities Face Challenges in Maintaining Complete and Accurate Data

As utilities continue to obtain and integrate data, they still face the challenge of maintaining complete and accurate data about their assets. The *International Infrastructure Management Manual* notes that data collection is a continuous process and that utilities need to remain consistent in gathering data and updating their central asset inventory as they repair, replace, or add infrastructure. Regular updating ensures that the information remains useful over time. To sustain the benefits garnered from its efforts to compile an accurate inventory, the Eastern Municipal Water District adopted a policy whereby employees must document changes to the inventory whenever assets are added, repaired, or removed. The utility has also developed methods to enforce its policy to make sure that the inventory is updated as required.

Interdepartmental Coordination and Information Sharing Present Difficult Cultural Challenges

According to industry officials, one of the major challenges to implementing asset management is changing the way utilities typically operate—in separate departments that do not regularly exchange information. It is essential to change this management culture, these officials believe, to encourage interdepartmental coordination and information sharing.

To encourage interdepartmental communication, utilities may have to train their employees in using the resources of other departments. For example, at the Orange County Sanitation District, the management team found it difficult to demonstrate to its employees that their job responsibilities do indeed affect the functions of the other departments. The utility's field staff possesses extensive information about the condition and performance of assets because they maintain these assets every day. However, these employees did not understand that the engineering department needs feedback on how the assets that the engineering department constructed are performing in the field. Such feedback could change future designs for these assets to improve their performance. As the utility implemented asset management, it established a work group to examine the conditions of asset failure, which provided a forum for the maintenance and engineering departments to collaborate. While this work group is still ongoing, one utility official noted that collaboration between these two departments will result in more efficient maintenance schedules for the utility's assets.

Similarly, the Eastern Municipal Water District reported that its middle-management team resisted some of the asset management changes because they believed these changes would limit their authority to manage their staff and workload. Before asset management, the utility maintained four different treatment facilities, each with its own maintenance staff. The utility believed that it could optimize its maintenance resources by combining all of the maintenance activities and staff at the four plants under one department. However, the managers at these treatment plants were reluctant to relinquish managerial control over their maintenance staff and feared that their equipment would be neglected. Once the new maintenance department was formed, however, these plant managers realized that centralizing these functions resulted in faster maintenance because the larger team could more effectively allocate time among the four facilities.

In some instances, utility employees may be reluctant to accept comprehensive asset management because it requires them to take on additional responsibilities when they are already pressed for time in their "day jobs." Additional time may indeed be necessary. According to officials at different utilities we visited, asset management requires staff throughout the organization to attend a variety of training programs— introductory, refresher, and targeted training by function or job—to ensure that they understand the value of asset management to both their own jobs and the operation of the utility.

Utilities' Efforts to Increase Focus on Long-Term Planning Conflict with Short-Term Priorities

While asset management provides utilities with information to justify needed rate increases, their justifications may not be effective because their governing body and their customers want to keep rates low. According to utility officials, governing bodies' reluctance to increase rates may be linked to constituent pressure to hold down user rates. In 2002, we reported that 29 percent of drinking water and 41 percent of wastewater utilities serving populations over 10,000 did not cover their full cost of service through user rates in their most recent fiscal year. Furthermore, about half of these utilities did not regularly increase their user rates; rather, they raised their user rates infrequently—once, twice, or not at all— from 1992 to 2001.

Utility officials and water industry organizations also note that utilities may have to respond to governing bodies' interests rather than to the long-term plan they developed using comprehensive asset management. For instance, while the Orange County Sanitation District's governing board has supported comprehensive asset management, it overrode utility plans for some capital projects and instead funded a $500 million secondary sewage treatment plant, which was not a utility priority. The board took this action in response to public concerns that the operating sewage plant was inadequate and had contaminated the water. A subsequent report showed, however, that the contamination more than likely did not result from an inadequate treatment plant. However, the utility will probably have to defer other priorities in order to design and build this new facility. In addition, the governing body may shift funding originally budgeted to implement the next phase of Orange County's asset management program to fund the new plant.

Several industry officials also pointed out that governing bodies for municipally owned utilities tend to make financial decisions about their drinking water and wastewater utilities in light of competing local needs that may be a higher priority for the electorate. One industry official also reported that locally elected officials tend to focus their efforts on short-term, more visible projects, while utility managers must focus on sustaining the utility's operation in the long term. For example, a utility's governing body may decide to forgo infrastructure repairs in order to build a new school or baseball field.

Smaller Utilities Can Benefit from Asset Management Despite Challenges Posed by Limited Resources

Smaller utilities can also benefit from the improved data, coordination, and informed decision making that result from asset management. Although small utilities represent a substantial portion of the water and wastewater industry,[26] officials recognize that these utilities may have more difficulty implementing asset management because they typically have fewer financial, technological, and staff resources. In addition, EPA has reported that small systems are less likely to cover their full cost of providing services because they have to spread their fixed infrastructure costs over a smaller customer base. However, EPA believes that comprehensive asset management will enable smaller systems to increase knowledge of their system, make more informed financial decisions, reduce emergency repairs, and set better priorities for rehabilitation and replacement.

Even the most rudimentary aspects of asset management can produce immediate benefits for small communities. For example, the Somersworth, New Hampshire, Department of Public Works and Utilities avoided a ruptured sewer main because it had collected data through its asset management initiative that mapped the location of critical pipelines.[27] As a result, when a resident applied for a construction permit to build a garage, the utility determined that one critical pipeline lay in the path of the proposed construction and could rupture. Therefore, the city of Somersworth denied the permit.

Similarly, the Department of Public Works in Denton, Maryland, which provides both drinking water and wastewater services, obtained positive results from applying asset management concepts without having to invest in sophisticated software or perform a complicated analysis. In this case, Denton's city council was apprehensive about investing in new trucks for the utility even though some of the existing trucks were in poor condition. Council members believed that it would be less expensive to continue repairing the existing fleet. However, using data collected through their asset management initiative, utility managers were able to track the maintenance and depreciation costs associated with these vehicles. As a result, they could demonstrate to their governing body that it was more cost-effective to purchase new vehicles than to continue repairing the older trucks.

Because smaller utilities have fewer capital assets to manage, industry officials noted that these utilities can implement asset management by turning to low-cost alternatives that do not require expensive or sophisticated technology. The small utilities can implement asset management using their existing asset data and recording this information in a central location that can be accessed by all of its employees, such as a set of index cards or an Excel spreadsheet. Similarly, the utility can adopt the practices of asset management incrementally, by initially making asset decisions based on their existing data.

SECTION 3: EPA CAN ENCOURAGE WATER UTILITIES TO USE ASSET MANAGEMENT BY STRENGTHENING EXISTING INITIATIVES

Opportunities exist for EPA to encourage water utilities' use of asset management by strengthening existing initiatives. Currently, EPA sponsors several initiatives to promote the use of asset management, such as training and informational materials, technical assistance, and research. While this is a good first step, the entities involved in these initiatives are not systematically sharing information within and across the drinking water and wastewater programs. With better coordination, however, EPA could leverage limited resources and reduce the potential for duplication within the agency. EPA could supplement its own efforts to disseminate information on asset management by taking advantage of similar efforts by other federal agencies, such as the Department of Transportation. Water industry officials also see a role for EPA in educating utility managers about how asset management can be a tool to help them meet regulatory requirements related to utility management. However, the officials raised concerns about the implications of mandating asset management as proposed in legislation being considered by the Congress.

EPA Sponsors Several Initiatives to Promote Utilities' Use of Asset Management

Through partnerships with water industry associations and universities, EPA has supported the development of training and informational materials to help drinking water and wastewater utilities implement asset management. In particular, EPA contributed funding toward the development of a comprehensive industry handbook on asset management, which was published in 2002 under a cooperative agreement with the Association of Metropolitan Sewerage Agencies.[1] The handbook lays out the principles of asset management and describes how utilities can use this approach to improve decision making, reduce costs, and ensure the longterm, high-level performance of their assets.

EPA has also sponsored materials specifically directed at small utilities. For small drinking water systems, EPA's Office of Ground Water and Drinking Water published a handbook in 2003 that describes the basic concepts of asset management and provides information on how to develop an asset management plan.[2] In addition, to help entities such as mobile home parks and homeowners' associations that own and operate their own water systems, the office is developing a booklet on preparing a simple inventory of the systems' assets and assessing their condition.[3] EPA's Office of Wastewater Management is funding the development of a "toolkit" by a university-based training center to help small wastewater utilities implement asset management. The toolkit is currently being field tested and is scheduled for release in 2006. Among other things, it includes self-audit instruments to help utility managers to analyze their systems' needs, training materials, and a summary of lessons learned in the field.

In addition to various informational materials on asset management, EPA has sponsored a number of training and technical assistance programs. For example, the Office of Wastewater Management, along with representatives from a major utility and an engineering firm, developed a 2-day seminar on asset management, which will be held at several locations around the country during fiscal year 2004. For smaller drinking water and wastewater utilities, EPA funds state and university-based centers that provide training and technical assistance to small utilities on a variety of matters, including asset management. Specifically

- EPA's Office of the Chief Financial Officer funds nine university-based "environmental finance centers" that assist local communities in seeking financing for environmental facilities, including municipal drinking water and wastewater utilities. In fiscal year 2003, the nine centers shared a total of $2 million in funding from the Office of the Chief Financial Officer; some centers also receive funds from EPA program offices for specific projects. According to an official in EPA's Office of Ground Water and Drinking Water, at least three of the finance centers have efforts related to asset management planned or underway to benefit drinking water utilities. For example, the centers at Boise State University and the University of Maryland provide on-site and classroom training on establishing an asset inventory; collecting data on the age, useful life, and value of capital assets; recordkeeping; financing; and setting rates high enough to cover the full cost of service. Regarding the latter topic, Boise State's finance center developed a simplified software program, called CAPFinance, which can help smaller systems collect and analyze the data they need in order to set adequate user rates; much of this information can be used to create a rudimentary asset management program.

- Another eight university-based technical assistance centers receive funding under the Safe Drinking Water Act to help ensure that small drinking water systems have the capacity they need to meet regulatory requirements and provide safe drinking water. In fiscal year 2003, the eight centers shared about $3.6 million in funding from the Office of Ground Water and Drinking Water. According to an official from that office, three of the centers are holding workshops or developing guidance manuals that focus on sustaining the financial viability of small systems in some way; the official believes that much of this material is relevant to implementing asset management.
- The Office of Wastewater Management funds 46 state and university-based environmental training centers under the Clean Water Act to train wastewater utility officials on financial management, operations and maintenance, and other topics. According to an official with EPA's wastewater program, one of the 46 centers is developing a series of six training courses to help small wastewater utilities implement some of the basic elements of asset management, such as inventorying system assets and assessing their condition.[4] Once this effort is completed, the center will disseminate the course materials to the remaining 45 centers so that staff from the other centers will be able to teach the asset management courses to operators of small wastewater utilities across the country.

EPA has also funded research projects related to asset management. For example, one project—sponsored by EPA, the Water Environment Federation, and the Association of Metropolitan Sewerage Agencies— examined the interrelationship between asset management and other management initiatives, such as environmental management systems, that have received some attention within the water industry.[5] The project found that to varying degrees, the initiatives share a common focus on continuous improvement through self-assessment, benchmarking, and the use of best practices and performance measures. The final report, issued in September 2002, concluded that while the initiatives overlap substantially, they are generally compatible.[6]

EPA also contributed $75,000 toward a 2002 report by the Water Environment Research Foundation, which summarized the results of a 2-day workshop held to develop a research agenda for asset management.[7] Workshop participants, who included utility managers, regulators, and industry consultants, identified areas in which they need improved tools and technical approaches, established criteria for evaluating asset management research needs, and identified and set priorities for specific research projects. According to the foundation's report, the workshop ultimately recommended 11 research projects, 2 of which will get underway in 2004. EPA is contributing $200,000 to one of these projects, which will develop protocols for assessing the condition and performance of infrastructure assets and predictive models for correlating the two. The foundation will fund the second project, which is scheduled to begin in March 2004, and will develop guidance on strategic planning for asset management. According to EPA, the second project will also develop a Web-based collection of best practices on asset management; utilities will be able to purchase licenses to gain access to the materials.

The remaining research projects identified in the workshop highlight the need for practical tools to help utilities implement the most fundamental aspects of asset management. They include projects to

- establish methodologies for determining asset value, compiling inventories, and capturing and compiling information on the assets' attributes;
- develop methodologies for calculating life-cycle costs for infrastructure assets;
- construct predictive models for infrastructure assets that project life-cycle costs and risks;
- identify best practices for operating and maintaining infrastructure assets by asset category, condition, and performance requirements; and
- identify best practices for integrating water and wastewater utility databases.

In addition, workshop participants recommended a project to assess the feasibility of establishing an Asset Management Standards Board for the drinking water and wastewater industry.

EPA's Efforts to Promote Asset Management Could Be Strengthened by Leveraging Ongoing Efforts Within and Outside the Agency

EPA could build on its efforts to promote asset management at drinking water and wastewater utilities by better coordinating ongoing and planned initiatives in the agency's drinking water and wastewater programs. In addition, EPA could leverage the efforts of other federal agencies, such as the Department of Transportation, that have more experience in promoting asset management as well as informational materials and tools that could potentially be useful as EPA and the water industry develop similar materials.

Improving Coordination within and across Drinking Water and Wastewater Programs Could Help Maximize Limited Resources

While some of EPA's efforts to promote the use of asset management, such as sponsoring the comprehensive industry handbook, have involved both the drinking water and wastewater communities, it appears that other efforts are occurring with little coordination between the drinking water and wastewater programs or other offices within EPA. For example, the Office of the Chief Financial Officer, the Office of Ground Water and Drinking Water, and the Office of Wastewater Management have funded parallel but separate efforts to develop handbooks, software, or other training materials to help small drinking water and wastewater utilities implement asset management or related activities such as improving financial viability. According to our interviews with EPA officials and representatives of the university-based training and technical assistance centers, no central repository exists for EPA to track what the university-based centers are doing and ensure that they have the information they need to avoid duplication and take advantage of related work done by others. The centers that share information do so primarily within their own network, as in the case of the environmental finance centers, or share information on an ad hoc basis. As a result, the centers are likely to miss some opportunities to exchange information. Similarly, the drinking water and wastewater program offices do not regularly exchange information on what they or their centers are doing to develop informational materials, training, or technical assistance on asset management.

EPA officials explained that, to some extent, the organizational framework within which the centers operate contributes to limited information sharing and duplication of effort. As a result, EPA is not maximizing the resources it devotes to encouraging utilities' use of asset management. In the case of the environmental finance centers, for example, each one negotiates a work plan with the EPA regional office it serves. Although EPA headquarters also has some influence over what the centers work on, the centers primarily focus on regional priorities and work with the states within the regional office's jurisdiction. Occasionally, EPA's drinking water and wastewater program offices fund projects at the environmental finance centers that are independent of their regional work plans. For example, the drinking water program provided some funds to the center at Boise State to develop an evaluation tool that states can use to assess utilities' qualifications for obtaining financial assistance from state revolving loan funds. For the most part, however, the training and technical assistance centers operate autonomously and do not have a formal mechanism for regularly exchanging information among the different center networks or between the drinking water and wastewater programs.

EPA Could Supplement Its Efforts to Promote Asset Management by Using Information Available from Other Federal Agencies

EPA has not taken advantage of the guidance, training, and implementation tools available from other federal agencies, which would help EPA leverage its resources. For the purposes of our review, we focused on the Department of Transportation's Federal Highway Administration because it has been involved in promoting asset management for about a decade and has been at the forefront of developing useful tools and training materials. In 1999, the Federal Highway Administration established an Office of Asset Management to develop tools and other materials on asset management and encourage state transportation agencies to adopt asset management programs and practices.

According to officials within the Office of Asset Management, the basic elements of asset management are the same regardless of the type of entity responsible for managing the assets or the type of assets being managed. Simply put, every organization needs to know the assets it has, their condition, how they are performing, and the costs and benefits of alternatives for managing the assets. Over the years, the Office of Asset Management has published several guidance documents on asset management and its basic elements. While the purpose of the guidance was to assist state transportation agencies, Transportation officials believe that the general principles contained in their publications are universally applicable. The office's guidance includes, for example,

- a general primer on the fundamental concepts of asset management;
- a primer on data integration that lays out the benefits of and tools for integrating data, the steps to follow in linking or combining large data files, potential obstacles to data integration and ways to overcome them, and experiences of agencies that have integrated their data;[8] and
- a primer on life-cycle cost analysis that provides information on how to apply this methodology for comparing investment alternatives and describes uncertainties regarding when and how to use life-cycle cost analysis and what assumptions should be made during the course of the analysis.[9]

Transportation's Office of Asset Management has also developed a software program to assist states in estimating how different levels of investment in highway maintenance will affect both user costs and the highways' future condition and performance. In addition, to disseminate information on asset management, the office established a Web site that includes its most recent tools and guidance and links to external Web sites with related asset management information, including a link to an asset management Web site jointly sponsored with the American Association of State Highway and Transportation Officials.

As EPA began its efforts to explore the potential of comprehensive asset management to help address utility infrastructure needs, officials from the Office of Water met with staff from Transportation's Office of Asset Management and obtained a detailed briefing on its asset management program. Although EPA officials expressed concerns about having relatively limited resources to promote asset management, they have so far not pursued a closer relationship with Transportation or other federal agencies with experience in the field. For example, EPA may find opportunities to adapt Transportation's guidance materials or use other efforts, such as a Web site that brings together asset management information from diverse sources, as a model for its own initiatives.

Water Industry Officials Favor an Expanded Role for EPA in Promoting Asset Management, but Raised Concerns About Additional Regulatory Requirements

Water industry officials support a greater role for EPA in promoting asset management, both as a tool for better managing infrastructure and for helping drinking water and wastewater utilities meet existing or proposed regulatory requirements. However, they stopped short of endorsing legislative proposals that would require utilities to develop and implement plans for maintaining, rehabilitating, and replacing capital assets, often as a condition of obtaining loans or other financial assistance.

Water Industry Officials See Role for EPA in Linking Asset Management to Regulatory Requirements and Other Initiatives Aimed at Improving Utility Management

To obtain views on the role that EPA might play in encouraging the use of asset management, we talked with officials from water industry associations and the 15 utilities that we selected for structured interviews. With few exceptions, the officials agreed that EPA should be promoting asset management in some way, although opinions varied on what activities would be most appropriate. One of the options that garnered the support of many was a greater leadership role for EPA in promoting the use of asset management. For example, 11 of the 15 utilities indicated that based on their own experience, asset management can help utilities comply with certain regulatory requirements that focus in whole or in part on the adequacy of utility infrastructure and the management practices that affect it. While EPA recognizes the link between asset management and regulatory compliance—and has noted the connection in some agency publications and training—some utility officials believe that EPA should increase its efforts in this regard. As examples of regulatory requirements for which asset management is particularly germane, officials from industry associations and individual utilities cited both the existing "capacity development" requirements under EPA's drinking water program and regulations for capacity, management, operation, and maintenance under consideration in the wastewater program, as follows:

- *Capacity development requirements for drinking water utilities.* To be eligible for full funding under the Safe Drinking Water Act's State Revolving Fund program, state regulatory agencies are required to have strategies to assist drinking water utilities in acquiring and maintaining the financial, managerial, and technical capacity to consistently provide safe drinking water. To assess capacity, states evaluate, among other things, the condition of the utilities' infrastructure, the adequacy of maintenance and capital improvement programs, and the adequacy of revenues from user rates to cover the full cost of service. Drinking water utilities that are determined to lack capacity are not eligible for financial assistance from the revolving loan fund.[10]

- *Capacity, management, operation, and maintenance requirements for wastewater utilities.* As part of its wastewater management program under the Clean Water Act, EPA is considering regulations designed to improve the performance of treatment facilities and protect the nation's collection system infrastructure by enhancing and maintaining system capacity (i.e., peak wastewater flows), reducing equipment and operational failures, and extending the life of sewage treatment equipment. Among other things, wastewater utilities would be required to prepare capacity, management, operation, and maintenance plans for their operations. The regulations would also require utilities to assess the condition of their physical infrastructure and determine which components need to be repaired or replaced.

According to industry officials, implementing asset management is consistent with meeting these requirements, and it enhances utilities' ability to comply with them. For the requirements being considered for wastewater utilities, for example, EPA has concluded that three basic components are a facility inventory, a condition assessment, and asset valuation—all of which are important elements of asset management. Consequently, the officials believe that it makes sense for EPA to place more emphasis on the use of comprehensive asset management.

Some water industry officials also told us that EPA should use the relationship between asset management practices and the financial reporting requirements under Governmental Accounting Standards Board Statement 34 as a means of promoting the use of asset management. Under these new requirements, state and local governments are required to report information about public infrastructure assets, including their drinking water and wastewater facilities. Specifically, the governments must either report depreciation of their capital assets or implement an asset management system.[11]

Given the infrastructure-related regulatory requirements and utilities' other concerns about the condition of their assets, it is not surprising that 11 of the 15 utilities we interviewed in depth saw a need for EPA to set up a clearinghouse of information on comprehensive asset management. Several utilities suggested that EPA establish a Web site that would serve as a central repository of such information. This site could provide drinking water and wastewater utilities with direct and easy access to information that would help them better manage their infrastructure. For example, the Web site could gather in one place the guidance manuals, tools, and training materials developed by EPA or funded through research grants and its training and technical assistance centers. The site could also contain links to asset management tools and guidance developed by domestic and international water associations or other federal agencies, such as Transportation's Office of Asset Management. Several officials also commented that it might be useful to have a site where drinking water and wastewater utilities could share lessons learned from implementing asset management. Other utilities also supported the idea of a Web site, but were uncertain about whether EPA was the appropriate place for it. In commenting on a draft of this report, EPA generally agreed that an EPA Web site devoted to asset management would be worthwhile and is considering developing such a site.

Water Industry Officials Cite Implementation Challenges if Asset Management Were to Be Mandated

In recent years, the Congress has considered several legislative proposals that would, in part, promote the use of asset management in some way. These proposals generally call for an inventory of existing capital assets; some type of plan for maintaining, repairing, and replacing the assets; and a plan for funding such activities. All but one of the proposals made having the plans a condition of obtaining federal financial assistance. The proposals are consistent with what we have found to be the leading practices in capital decision making. As we reported in 1998, for example, routinely assessing the condition of assets allows managers to evaluate the capabilities of existing assets, plan for future replacements, and calculate the cost of deferred maintenance.[12] However, according to key stakeholders, implementing and enforcing requirements for asset management could be problematic at this time.

We asked water industry groups, associations of state regulators, and individual utilities for their views on the proposed mandate of asset management plans. While most of them endorse asset management, they raised several concerns about a statutory requirement. For example:

- Officials from water industry associations believe that drinking water and wastewater utilities are already overburdened by existing regulatory requirements and that many utilities lack the resources to meet an additional requirement for developing asset management plans.

- The Association of State Drinking Water Administrators and the Association of State and Interstate Water Pollution Control Administrators both said that the states lack the resources to oversee compliance and determine the adequacy of asset management plans. Both the state and industry associations questioned the feasibility of defining what would constitute an adequate plan.
- Officials at 12 of the 15 utilities where we conducted in-depth interviews had serious reservations about a requirement. For example, some utility managers were concerned that EPA and the states would attempt to standardize asset management and limit the flexibility that utilities need to tailor asset management to their own circumstances. Another concern was that the states lack financial and technical resources and thus are ill equipped to determine whether utilities' asset management plans are adequate. Finally, some utility officials also questioned the burden that such a requirement would place on small utilities.

Other utility officials either support a requirement or support the concept of asset management but question whether mandating such a requirement is an appropriate role for the federal government. One of the officials commented that whether or not asset management is required, utilities should manage their infrastructure responsibly and charge rates sufficient to cover the full cost of service. The National Association of Water Companies, which represents investor-owned utilities, supports a requirement for asset management to ensure that public water and wastewater utilities are operating efficiently and are charging rates that cover the full cost of service.

Conclusion

Comprehensive asset management shows real promise as a tool to help drinking water and wastewater utilities better identify and manage their infrastructure needs. Even with their limited experience to date, water utilities reported that they are already achieving significant benefits from asset management. EPA clearly recognizes the potential of this management tool to help ensure a sustainable water infrastructure and has sponsored a number of initiatives to support the development of informational materials and encourage the use of asset management. However, in an era of limited resources, it is particularly important for EPA to get the most out of its investments by coordinating all of the asset management-related activities sponsored by the agency and taking advantage of tools and training materials developed by others—including domestic and international industry associations and other federal agencies with experience in asset management.

Establishing a central repository of all asset management-related activities could not only foster more systematic information sharing but also help minimize the potential for duplication and allow EPA-sponsored training and technical assistance centers to build on each other's efforts. As EPA has recognized, improving utilities' ability to manage their infrastructure cannot help but improve their ability to meet regulatory requirements that focus on the adequacy of utility infrastructure and management practices. Consequently, it is in the agency's best interest to disseminate information on asset management and promote its use. Establishing a Web site, perhaps as part of the repository, would help ensure that such information is accessible to water utilities and that EPA is getting the most use out of the materials whose development it funded. Moreover, EPA could use the site as a means of strengthening its efforts to educate utility managers on the connection between effectively managing capital assets and the ability to comply with relevant requirements under the Safe Drinking Water Act and Clean Water Act.

Recommendations for Executive Action

Given the potential of comprehensive asset management to help water utilities better identify and manage their infrastructure needs, the Administrator, EPA, should take steps to strengthen the agency's existing initiatives on asset management and ensure that relevant information is accessible to those who need it. Specifically, the Administrator should

- better coordinate ongoing and planned initiatives to promote comprehensive asset management within and across the drinking water and wastewater programs to leverage limited resources and reduce the potential for duplication;
- explore opportunities to take advantage of asset management tools and informational materials developed by other federal agencies;
- strengthen efforts to educate utilities on how implementing asset management can help them comply with certain regulatory

requirements that focus in whole or in part on the adequacy of utility infrastructure and the management practices that affect it; and

- establish a Web site to provide a central repository of information on comprehensive asset management so that drinking water and wastewater utilities have direct and easy access to information that will help them better manage their infrastructure.

APPENDIX I: UTILITIES SELECTED FOR STRUCTURED INTERVIEWS

Augusta County Service Authority, Verona, Virginia
Charleston Commissioners of Public Works, Charleston, South Carolina
Greater Cincinnati Water Works, Cincinnati, Ohio
Denton Department of Public Works, Denton, Maryland
Des Moines Water Works, Des Moines, Iowa
East Bay Municipal Utility District, Oakland, California
Eastern Municipal Water District, Perris, California
Louisville Water Company, Louisville, Kentucky
Massachusetts Water Resources Authority, Boston, Massachusetts
Orange County Sanitation District, Fountain Valley, California
Sacramento Regional County Sanitation District, Mather, California
Seattle Public Utilities, Seattle, Washington
Somersworth Department of Public Works and Utilities, Somersworth, New Hampshire
Tacoma Water, Tacoma, Washington
Western Carolina Regional Sewer Authority, Greenville, South Carolina

APPENDIX II: GAO CONTACTS IN AUSTRALIA AND NEW ZEALAND

Department of Treasury and Finance Government agency	Victoria State or region	Australia Country
Essential Services Commission	Victoria	Australia
Independent Pricing and Regulatory Tribunal	New South Wales	Australia
Manawatu District Council	Manawatu-Wanganui	New Zealand

Department of Treasury and Finance Government agency	Victoria State or region	Australia Country
Queensland Audit Office	Queensland	Australia
Citiwater Utility	Queensland State or region	Australia Country
Hunter Water Corporation	New South Wales	Australia
Melbourne Water Corporation	Victoria	Australia
South Australian Water Corporation	South Australia	Australia
Sydney Water Corporation	New South Wales	Australia
Watercare Services Limited	Auckland	New Zealand

End Notes for Executive Summary

[1] See U.S. General Accounting Office, *Water Infrastructure: Information on Financing, Capital Planning, and Privatization*, GAO-02-764 (Washington, D.C.: Aug. 16, 2002).

[2] See U.S. General Accounting Office, *Executive Guide: Leading Practices in Capital Decision-Making*, GAO/AIMD-99-32 (Washington, D.C.: December 1998).

End Notes for Section 1

[1] See U.S. General Accounting Office, *Water Infrastructure: Information on Federal and State Financial Assistance*, GAO-02-134 (Washington, D.C.: Nov. 30, 2001). We adjusted the dollar amounts from the report to constant 2003 dollars.

[2] From fiscal year 2001 to fiscal year 2004, EPA provided $8.8 billion for water infrastructure and Agriculture provided $6 billion, in constant 2003 dollars.

[3] The Clean Water State Revolving Fund may also be used for other water quality improvement projects, such as nonpoint source pollution control and estuary management, in addition to wastewater treatment facilities.

[4] Simply stated, the funding gap is equal to the estimated infrastructure needs less projected spending.

[5] U.S. Environmental Protection Agency, *Drinking Water Infrastructure Needs Survey: Second Report to Congress*, EPA 816-R-01-004 (Washington, D.C.: February 2001) and U.S. Environmental Protection Agency, *Clean Watersheds Needs Survey 2000: Report to Congress*, EPA 832-R-03-001 (Washington, D.C.: August 2003).

[6] See GAO-02-764, 3. We sent questionnaires to 1,425 drinking water systems and 2,391 wastewater systems serving more than 10,000 people. In our analysis, utilities were weighted to account statistically for all utilities serving populations greater than 10,000, including those not selected for our sample.

[7] American Water Works Association Water Industry Technical Action Fund, *Dawn of the Replacement Era: Reinvesting in Drinking Water Infrastructure* (Denver, Colo.: May 2001).

[8] Water Environment Research Foundation, *New Pipes for Old: A Study of Recent Advances in Sewer Pipe Materials and Technology* (2000), 4-1.

[9] GAO-02-764, 7, 35, 42.

[10] We do not have specific information on the fiscal years covered in utilities' responses; however, we sent our survey out during September 2001.

[11] For the purposes of our survey, we focused on the asset planning elements identified by the Governmental Accounting Standards Board in a June 30, 1999, statement that made comprehensive changes in state and local governments' financial reporting requirements. Among other things, it requires, for the first time, the governments to report information about public infrastructure assets, including their drinking water and wastewater facilities. Specifically, the governments must begin reporting depreciation of their capital assets or implement an asset management system. See *Governmental Accounting Standards Board Statement No. 34, Basic Financial Statements—and Management's Discussion and Analysis—for State and Local Governments.*

[12] For example, see (1) *Managing Public Infrastructure Assets to Minimize Cost and Maximize Performance*, developed by the Association of Metropolitan Sewerage Agencies in partnership with the American Water Works Association, the Association of Metropolitan Water Agencies, and the Water Environment Federation (2002); (2) *International Infrastructure Management Manual*, published jointly by the Association of Local Government Engineering New Zealand, Inc., and the Institute of Public Works Engineering of Australia (2002); (3) *Asset Management Primer*, published by the U.S. Department of Transportation (December 1999); and (4) *Executive Guide: Leading Practices in Capital Decision-Making*, published by GAO (GAO/AIMD-99-32; December 1998).

[13] For example, in the case of a drinking water utility, a water main serving the local hospital might be identified as a critical asset because the consequences of a water main break could be significant.

[14] See *International Infrastructure Management Manual*, published jointly by the Association of Local Government Engineering New Zealand, Inc., and the Institute of Public Works Engineering of Australia (2002).

[15] We were interested in learning about the applicability of asset management at small utilities because they represent the vast majority of the regulated systems, with 93 percent of community drinking water utilities and 71 percent of wastewater utilities serving 10,000 people or fewer.

End Notes for Section 2

[1] The Massachusetts Water Resources Authority provides water and sewer services to approximately 2.5 million people.

[2] Seattle Public Utilities is a utility serving approximately 1.3 million drinking water customers and about 500,000 wastewater customers.

[3] The Louisville Water Company provides water services to approximately 800,000 people in the Louisville, Kentucky, area and in parts of Oldham and Bullitt counties.

[4] The Sacramento Regional County Sanitation District is a California wastewater utility serving approximately 482,000 customers in the Sacramento area.

[5] Eastern Municipal Water District is a water and wastewater utility serving approximately 501,000 customers in Southern California.

[6] The South Australian Water Corporation, an Australian utility located in the state of South Australia, provides water and wastewater services for approximately 1.4 million people.

[7] Saco, Maine, is a city of approximately 16,800 people. Among other services, the city government is responsible for providing wastewater treatment. The city asked to have its bond rating reassessed before beginning a school renovation project.

[8] Western Carolina Regional Sewer Authority provides wastewater treatment services to approximately 360,000 customers in Greenville County, South Carolina, and portions of Spartanburg, Laurens, and Anderson counties.

[9] The Charleston Commissioners of Public Works provides water and wastewater services to more than 400,000 customers in the Charleston, South Carolina, area.

[10] These regulations are under consideration. EPA proposed the regulations in January 2001, but in accordance with the incoming administration's regulatory review plan, withdrew the proposal to give the administration an opportunity to review it.

[11] The Orange County Sanitation District provides wastewater services for approximately 2.3 million people living in central and northwest Orange County, California.

[12] The operating cost reductions were measured per property, or for each residential or community property connected to the water and sewer supply.

[13] Hunter Water Corporation is an Australian utility that provides water and wastewater services to almost 500,000 people in parts of New South Wales.

[14] Hunter Water's capital expenditures fluctuated during the 10-year period, decreasing from about $40 million in 1990/1991 to about $10 million in 1997/1998, then spiking to about $62 million in 2000/2001, then decreasing to about $40 million again. Utility officials attribute the spike in capital expenditures to growth and a regulatory upgrade of the utility's wastewater treatment system and transport system.

[15] An environmental management system is a management tool to help an organization improve its environmental performance, prevent pollution, and meet regulatory requirements.

[16] EPA used these mislabeled Australian estimates as a basis for projecting life-cycle cost savings of 20 to 30 percent for U.S. utilities using asset management. Additionally, it appears that the engineering firm that predicted about $350 million in life-cycle cost savings for Orange County Sanitation District used the same estimates in its model.

[17] Augusta County Service Authority serves a population of approximately 67,000 people in Virginia.

[18] East Bay Municipal Utility District supplies water and provides wastewater treatment to approximately 1.3 million drinking water customers and 640,000 wastewater customers in parts of Alameda and Contra Costa counties in northern California.

[19] Tacoma Water serves approximately 300,000 customers in the city of Tacoma, Washington, and portions of Pierce and South King counties.

[20] The Department of Public Works in Billerica, Massachusetts, serves a population of approximately 10,000 people.

[21] Des Moines Water Works distributes water to a population of approximately 300,000 in Des Moines, Iowa, and its surrounding communities.

[22] See *Managing Public Infrastructure Assets to Minimize Cost and Maximize Performance*, 57.

[23] Greater Cincinnati Water Works provides water to approximately 1.2 million customers.

[24] Melbourne Water Corporation manages the city of Melbourne's water catchments and major distribution system. The utility supplies approximately 500,000 megaliters of water annually to its three retail water companies.

[25] St. Paul Regional Water Services provides drinking water to approximately 415,000 residents of St. Paul, Minnesota, and its surrounding communities.

[26] As noted earlier, most U.S. utilities are small, with 93 percent of the 54,000 community drinking water systems and 71 percent of the 16,000 wastewater systems serving 10,000 people or fewer.

[27] Somersworth is a New Hampshire city with approximately 11,000 residents. Somersworth's Department of Public Works and Utilities is responsible for water and wastewater services for the city.

End Notes for Section 3

[1] The Association of Metropolitan Sewerage Agencies developed the handbook, *Managing Public Infrastructure Assets to Minimize Cost and Maximize Performance,* in partnership with the American Water Works Association, the Association of Metropolitan Water Agencies, and the Water Environment Federation.

[2] U.S. Environmental Protection Agency, *Asset Management: A Handbook for Small Water Systems*, EPA 816-R-03-016 (Washington, D.C.: September 2003).

[3] EPA has drafted the booklet, *Taking Stock of Your Water System*, and expects to publish the final version sometime in 2004.

[4] The training courses are being developed by the environmental training center located at the College of Southern Maryland, which is also responsible for the asset management toolkit for small wastewater utilities. To fund this work, EPA awarded the center a 3-year grant totaling $450,000, covering the period from August 2002 to August 2005.

[5] An environmental management system is a management tool to help an organization improve its environmental performance, prevent pollution, and meet regulatory requirements.

[6] Water Environment Federation and Association of Metropolitan Sewerage Agencies, *Moving Toward Comprehensive Utility Management Systems: Report of the Environmental Management Systems (EMS) for Public Utilities Integration Project* (Alexandria, Va.; Washington, D.C.: September 2002). EPA provided a grant of $256,888 for this project.

[7] See Water Environment Research Foundation, *Research Priorities for Successful Asset Management: A Workshop* (Alexandria, Va.: 2002).

[8] U.S. Department of Transportation, Federal Highway Administration, Office of Asset Management, *Data Integration Primer* (Washington, D.C.: August 2001).

[9] U.S. Department of Transportation, Federal Highway Administration, Office of Asset Management, *Life-Cycle Cost Analysis Primer* (Washington, D.C.: August 2002).

[10] States may nevertheless provide financial assistance if the use of such assistance will ensure compliance with regulatory requirements and the water utility has agreed to make the necessary changes in operations to ensure that it has the financial, managerial, and technical capacity to comply over the long term.

[11] Privately owned utilities are not required to comply with financial reporting requirements from the Governmental Accounting Standards Board. About half of the nation's drinking water systems and an estimated 20 percent of the wastewater systems are privately owned, according to EPA and industry sources.

[12] GAO/AIMD-99-32, 26.

INDEX

A

B

C

D

E

F

G

H

I

L

M

N

O

P

Q

R

S

T

U

W